FOCUS ON CIVILIZATIONS AND CULTURES – MUSIC

MUSIC: COMPOSITION, INTERPRETATION AND EFFECTS

Focus on Civilizations and Cultures – Music

The Brest Boys' and Young Men's Choir (A CD-ROM Production)
Alla Igumnova (Editor)
2003. ISBN: 1-59033-670-4

African Music: Traditional and Contemporary
Alexander Akorlie Agordoh
2005. ISBN: 1-59454-554-5

Musical ConverSings with Children
Linda Neelly
2007. ISBN: 1-60021-708-7

The Vano Saradjishvili Tbilisi State Conservatoire 1917-2007
Manana Doijashvili
2008. ISBN: 1-60021-910-1

Echoes from Georgia: Seventeen Arguments on Georgian Polyphony
Rusudan Tsurtsumia and Joseph Jordania (Editors)
2010. ISBN: 978-1-60876-477-8

Music: Composition, Interpretation and Effects
Tamara E. Ivanova (Editor)
2010. ISBN: 978-1-60876-170-8

FOCUS ON CIVILIZATIONS AND CULTURES – MUSIC

MUSIC: COMPOSITION, INTERPRETATION AND EFFECTS

TAMARA E. IVANOVA
EDITOR

Nova Science Publishers, Inc.
New York

For permission to use material from this book please contact us:
Telephone 631-231-7269; Fax 631-231-8175
Web Site: http://www.novapublishers.com

LIBRARY OF CONGRESS CATALOGING-IN-PUBLICATION DATA

Music : competition, interpretation, and effects / editors [sic], Tamara E. Ivanova.
p. cm.
Includes bibliographical references and index.
ISBN 978-1-60876-170-8 (hardcover : alk. paper)
1. Music--Psychological aspects. 2. Music--Social aspects. I. Ivanova, Tamara E.
ML3830.M973 2009
781'.11--dc22

2009031997

Published by Nova Science Publishers, Inc. ✦ New York

CONTENTS

PREFACE

Music is an art form whose medium is sound organized in time. Common elements of music are pitch (which governs melody and harmony), rhythm (and its associated concepts tempo, meter, and articulation), dynamics, and the sonic qualities of timbre and texture. The creation, performance, significance, and even the definition of music vary according to culture and social context. Music can be divided into genres and subgenres, although the dividing lines and relationships between music genres are often subtle, sometimes open to individual interpretation, and occasionally controversial. Within "the arts", music may be classified as a performing art, a fine art, and auditory art. To people in many cultures, music is inextricably intertwined into their way of life. Music is known to have many benefits for health, ranging from inducing relaxation and sleep to improving a depressed mood. Music also enhances cognitive function such as attention to task, communication, learning and memory. This new and important book gathers the latest research from around the globe in this dynamic field of study and highlights topics such as: rap music and its effect on sexual attitudes, music videos and risky driving, burnout and music use, music memory and the right hippocampus, and others.

In Chapter 1, the authors review the recent literature on music and social influence, with a particular emphasis on the influence of music on sexual attitudes and behaviors. The authors also present an experimental study that provided an empirical investigation of the relationships between sexually explicit rap music and sexual attitudes, perceived norms, and behaviors. Results indicated no significant short-term effects of rap music on the expression of sexual attitudes or perception of norms. However, the authors found marginal support for sexually explicit rap in priming condom-related sexual behavior.

In Chapter 2 the authors review their understanding of musical emotion focusing in the hypotheses, approaches, and results obtained in their interdisciplinary group since 2001 when an initial study proved that different musical excerpts evoke a significant inter-observer agreement in the selection of previously organized emotion terms within a relatively homogeneous population of human subjects. The authors have used both a correlation measure among the different EEG electrode sites and metabolic images of brain function in non-musician human volunteers during the hearing of three instrumental music masterpieces selected for their capacity to generate pleasant or unpleasant music emotions. Using both fMRI and EEG coherence brain images during the listening of the three masterpieces, the authors found that music-induced emotions are sustained by an activation of specific brain sites and changes in the pattern of coherent activity between cortical regions. Both brain imaging approaches identified divergent brain networks for positive and negative music affective valence. A left cortical system involved with pleasant music feelings included the posterior temporal-parietal, occipital, and middle prefrontal regions. A coherent activity in upper alpha frequencies took place between most left hemisphere electrodes with a focus in the temporo-parietal and occipital regions. In other studies the authors found that the coherent alpha activity of the left hemisphere involved in pleasant music emotion is much larger in women linking all frontal regions with the left temporal and parietal-temporal association areas of the brain. Even though women and men reported similar levels of all emotions, the self-reported feeling of being happy was only significant in women and the number of significant adjectives of the *happy* component was more abundant in them. The predominance of left hemisphere activation with pleasant musical feelings and right activation with unpleasant ones is consistent with right hemisphere predominance for dealing with novel cognitive situations and the left for predictable representations. The emotional effects of music may have to do with symbolic and grammatical properties that engage representational semantic systems of the mind/brain. The activations of extensive networks gain emotional meaning in higher neocortical substrata intersecting with the language systems of the brain. The affective powers of music require higher symbolic mind/brain mechanisms developed and functioning in unison with cultural cosmologies. Music expresses and carries emotive and symbolic meanings capable of extending and transcending natural stimuli by impinging into deeply meaningful human emotions engaging semantic neocortical neuron networks.

Purpose: Music video viewing is a very popular pastime among adolescents. The public criticism of music videos has encouraged researchers to examine the effects of music video exposure. Although an association between music video exposure and several health risk behaviors (e.g. drinking) has been found, the relationship between music video viewing and risky driving has remained largely unexamined. In Chapter 3 the relationship between adolescents' music video viewing and risky driving such as driving after consuming alcohol and joy riding is explored.

Methods: Participants were 354 adolescent males and females who participated in a panel study (2-year interval). Respondents were 17 or 18 years old during the first wave of data collection and did not have their driver's license yet. They completed a questionnaire on music video viewing, sensation seeking, aggression, attitudes towards joy riding and driving after the consumption of alcohol, and the intention to perform these behaviors in the future. Two years later the respondents had obtained their driver's license and were questioned on their actual risky driving behavior. The relationships between these constructs were analyzed using structural equation models.

Results: The results indicate that music video viewing is indirectly associated with joy riding and driving after the consumption of alcohol through the attitudes towards these behaviors and the intention to perform these behaviors in the future. More music video viewing resulted in a more positive attitude towards risky driving, even after controlling for sensation seeking and aggression. These attitudes are positively related to the intention to perform these reckless behaviors in the future and these intentions are, in turn, a good predictor of the actual risky driving behavior two years later.

Conclusions: The results indicate that music video viewing during adolescence is an important and significant predictor of risky driving two years later.

Music is known to have many benefits for health, ranging from inducing relaxation and sleep to improving a depressed mood. Music also enhances cognitive function such as attention to task, communication, learning, and memory. The objective of Chapter 4 is to describe several empirically verified health strategies that are based on music. These strategies have been used across several clinical trials to assist individuals with managing serious chronic illnesses and to improve their health and safety.

There is stress in the computer information systems industry that affects work performance and employee wellness [1, 2, 3]. The nature of the business of creating information systems is competitive and frequently involves aggressive deadlines. The designers and developers of these systems are required to be highly innovative and efficient in the face of constant time pressures. The leading consequence of the stressors is reported as increased frustration which "negatively impacts a job that requires high levels of discipline, focus, and systematic thinking" [1, p. 73].

In contrast, evidence from psychological literature suggests that mild positive moods contribute to creative problem solving in task performance [4]. Increases in positive mood helped explain the improved work performance of computer information systems developers (CISD) when working with preferred music [5]. In a recent study examining the prevalence of personality type and music use [6] in two different CISD groups, there were twice as many individuals with introversion than with extraversion; a trend opposite to the general population of nearly a two to one ratio of extraverts to introverts. Those with introversion also had significantly greater negative mood than those with extraversion, with possible interference to optimal work performance. Introverts, more so than extraverts, are reported to make use of music for emotional regulation [7]. Thus, there is a need to account for the way in which music influences these individual differences and consequential work performance.

Greater evidence of the beneficial use of music in this high-stress occupation is needed and should be addressed by a) obtaining an objective productivity score, as compared to previously self-reported productivity measures, and b) accounting for differential music effects on productivity for extraversion and introversion, and c) accounting for the effect of a timed music-condition for work performance.

Research from organization science is currently emphasizing the role of daily moods in job behaviors, job satisfaction, and work outcomes [8], as well as the importance of enhancing self-adaptation, referred to as "thriving", in the workplace. [9]. Additionally, differences in personality type will determine differential mood and arousal needs and consequential work outcomes [10]. Chapter 5 contributes to the research on affect in the organizational literature, and, is pioneering in music-organizational literature for music psychology and music therapy. Lastly, the findings have practical implications for management of high-cognitive demand, high-stress occupations.

Social scientists have used a variety of methods to study the personalities of fans of various music genres. Some psychologists and sociologists have performed in-depth studies that have included participant observation, interviews and the administration of paper and pencil surveys and personality profiles (for example, Arnett, 1996, studied fans of heavy metal music, and Adams and Sardiello, 2000, studied fans of the Grateful Dead). More recently, Rentfrow and Gosling (2003) developed the Short Test of Music Preferences (STOMP) which they have administered to large samples of undergraduates at the University of Texas, along with various assessments of the Big Five personality traits. They concluded that music preferences fall within one of four distinct clusters (reflective and complex music, intense and rebellious music, upbeat and conventional music, energetic and rhythmic music) and that those who give high ratings to each of the four have distinct personality patterns.

Zweigenhaft (2008), using an updated version of the STOMP, and a more in-depth measure of the Big Five personality traits, found some support for the patterns in Rentfrow and Gosling's (2003) studies, but also found important differences in the personality profiles of those who liked certain of the music genres. He also found some clear differences between fans of different music genres that fell within the clusters that Rentfrow and Gosling (2003) identified. The study described in Chapter 6 seeks to look in greater depth at the personality patterns for fans of two kinds of music who are frequently characterized based on limited data, or are stereotyped based on anecdotal data: those who like heavy metal music and those who like country music. The findings indicate that although there are kernels of truth in the previous portrayals social scientists have made about the fans of these music genres, they are also inaccurate in some meaningful ways.

Although it has been established that the hippocampus plays a crucial role in verbal episodic memory, its role in music memory remains unknown. In Chapter 7 the authors introduce the recent findings on the relationship between music memory and the right hippocampus. First, the authors introduce their previous study which demonstrated that the right hippocampus selectively responded to retrieval success of music memory. The study also demonstrated a clear functional dissociation between the left and right hippocampi in successful retrieval of music. Second, the authors briefly review studies that showed the recruitment of the right hippocampus related to cognitive processing of music. Many of those studies were conducted with patients suffering from brain lesions. Because there were very few cases in which the brain lesions were limited to the hippocampus, it is difficult to identify the precise role of the hippocampus. Because systematic brain imaging studies have just started, it is premature to conclude the role of the hippocampus in music processing. Third, the authors review studies that showed hippocampal activity related to successful retrieval of verbal memory. Some schemas have been proposed about the role of the hippocampus in memory. They are not independent of each other, and may reflect several aspects of the function of the hippocampus. The role of the hippocampus may be too complex to be explained by a single schema. The authors hope that their experiment of retrieval success of music will contribute to better understanding hippocampal functions.

As explained in Chapter 8, functional cerebral areas involved in listening to emotionally charged music are wider than those activated by neutral music. Based on the pathophysiology of musicogenic epilepsy, a rare form of complex reflex epilepsy in which seizures are triggered by music, the authors can infer the functional organization of the brain's processing of music. The authors studied a 36-year-old, right-handed male amateur musician, who has had weekly epileptic partial seizures every time he listened to or played music with a strong emotional charge since the age of 24 years. The patient underwent prolonged video-polygraphic recording which documented three right temporal seizures preceded for several seconds by an increase in heart rate and blood pressure, all triggered by listening to music with an emotional content. During functional neuroimaging (fMRI) he listened to both "neutral" and "emotionally charged" music. The "neutral music" activated only a small eloquent area in the right temporal lobe (acoustic area), whereas an "emotionally charged melody" activated diffuse eloquent areas on the fronto-temporo-occipital lobes of the right hemisphere before seizure onset. Numerous studies have demonstrated the predominant involvement of right hemisphere structures in networks involved in processing musical information. Most cases of musicogenic epilepsy showed the dominant role of the right temporal lobe disclosed by functional neuroimaging and neurophysiological techniques. The triggering stimulus in their patient seems to be a strong emotional feeling induced by specific melodies, as his seizures were preceded by an increase in heart rate and blood pressure. Their results emphasize the role of the right temporal lobe in musicogenic epilepsy and show that the cerebral areas activated during the emotional status leading to seizures encompass the auditory cortex activated by neutral music.

Source Separationmethods applied tomusical signals have been shown to provide very interesting results for music processing. In this context, music unmixing, automatic music transcription and source-oriented reprocessing is possible using advanced digital signal processing techniques. Most musical signals are mixtures of several sources (mainly instruments and singers) which are active simultaneously. Sound Source Separation (SSS) aims at recovering each source signal from a set of mixture signals of the original sources, such as those obtained by a stereo microphone pair or an audio CD. Although some methods are able to provide a solution to the SSS problem, they are limited to certain mixing scenarios or the resulting separation quality is not as good as expected, which may difficult their application to practical music processing systems. In fact, the objective of SSS algorithms is usually to separate with the highest quality, but it is also possible to perform separation just to facilitate a posterior semantic analysis of complex audio material. Thus, the objective of SSS methods may be completely different depending on the application. In Chapter 9, the authors review some of the existing techniques for sound source separation and the authors discuss their practical use for musical applications.

In Chapter 10, a cocktail digital watermarking scheme based on echo effect is proposed to embed copyright information into music clips before distribution and to retrieve this information later to prove the ownership. The proposed scheme takes advantage of stereo music to increase retrieval capacity and robustness by embedding data into the left and the right channel using low-bit coding and echo hiding method which have complimentary characteristics against various attacks. Furthermore, an exponential time-spread echo kernel is proposed for echo hiding to embed data as echoes into music clips while preserving their sound natural. By spreading the echo so that it seems like natural room reverberation, through the use of two channels, the proposed scheme provides satisfactory sound quality and exhibits both efficient retrieval capacity and robustness against attacks. Experiments show that the proposed system retrieves watermarks with 100% accuracy for all clips. The embedding transparency of the proposed scheme is -1.28 in Objective Different Grade (ODG) which is equal to "almost imperceptible". With a set of 42 attacks supplied by Stirmark Benchmark for Audio (SMBA), the proposed scheme has an average robustness of 0.98 where the ideal value is 1.00, and higher than the conventional methods.It also shows the robustness against mp3 compression from 192kbps to 128 kbps. With these results, the proposed scheme is shown to be applicable to music distribution.

Musical Expressivity can be defined as the deviation from a musical standard when a score is performed by a musician. This deviation is made in terms of intrinsic note attributes like pitch, timbre, timing and dynamics. The advances in computational power capabilities and digital sound synthesis have allowed real-time control of synthesized sounds. Expressive control becomes then an area of great interest in the sound and music computing field. Musical expressivity can be approached from different perspectives. One approach is the musicological analysis of music and the study of the different stylistic schools. Chapter 11 provides a valuable understanding about musical expressivity. Another perspective is the computational modelling of music performance by means of automatic analysis of recordings.

It is known that music performance is a complex activity that involves complementary aspects from other disciplines such as psychology and acoustics. It requires creativity and eventually, some manual abilities, being a hard task even for humans. Therefore, using machines appears as a very interesting and fascinating issue. In this chapter, the authors present an overall view of the works many researchers have done so far in the field of Expressive Music Performance, with special attention to the computational approach.

In: Music: Composition, Interpretation and Effects
Editor: Tamara E. Ivanova, pp. 1-29
ISBN: 978-1-60876-170-8

Chapter 1

THE EFFECT OF SEXUALLY EXPLICIT RAP MUSIC ON SEXUAL ATTITUDES, NORMS, AND BEHAVIORS

Lucrezia M. Alcorn and Anthony F. Lemieux
Purchase College, State University of New York

Abstract

In this chapter, we review the recent literature on music and social influence, with a particular emphasis on the influence of music on sexual attitudes and behaviors. We also present an experimental study that provided an empirical investigation of the relationships between sexually explicit rap music and sexual attitudes, perceived norms, and behaviors. Results indicated no significant short-term effects of rap music on the expression of sexual attitudes or perception of norms. However, we found marginal support for sexually explicit rap in priming condom-related sexual behavior.

Introduction

The relationship between sexually explicit rap music and its influence on sexual behaviors has been widely assumed (e.g., Gore; Steinem; the American Academy of Pediatrics, the PTA, the National Coalition on Television Violence, and the Parents' Media Resource Center, as discussed in Hansen, 1995) although little empirical research has directly addressed this issue. Proponents of censorship base their arguments primarily on the notion that rap music features content of seemingly dubious moral standing, without providing empirical evidence that exposure to this type of music actually causes deleterious effects among listeners. Rap has been a particular target, and its criticisms have included both social and aesthetic aspects (see Lemieux, 2000 for a detailed discussion). Because of the substantial amount of negative attention that rap has generated, it is important to understand the biases inherent in its evaluation (see North and Hargreaves, 2008 for a detailed review). The fundamental logic of the campaigns to censor or 'sanitize' rap music, in particular, has been to assume that the messages in the music are particularly influential among its listeners. In this chapter, we present research that tests the core of this claim by focusing on the potential for sexually explicit rap music to influence attitudes, perceived social norms, and behaviors related to sexuality.

A complex issue raised by the focus on rap music is the assumption that it exists in a vacuum—where it may be *the* primary influence. However, in reality, music is just one component of a larger multimedia market that is rapidly expanding (North and Hargreaves, 2008). And in downplaying the importance of the broader social and cultural context, pro-censorship campaigners seem to disregard the irony, satire, and witticisms that are often present in rap music. Following from this, it seems that the most ardent critics of rap focus on the explicit "sex and drugs" content of rap music and presume that these are at the root of ongoing social problems. As mentioned earlier, there is a limited amount of experimental evidence available to support the notion that this music *causes* changes in sexual attitudes and behaviors, and it is not clear which aspects of the media cause the effect. Part of the problem is that it is very difficult, if not impossible, to publish null results in scientific journals, which means that unpublished data could exist that shows non-significant or no effects of rap music on sexual attitudes and behaviors (North and Hargreaves, 2008).

It stands to reason that if listening to rap music were to cause changes in attitude and behavior irrespective of other potential mediating factors, then we would expect to see these effects among all rap music fans. The research that does exist appears to indicate that specific individual differences among listeners, such as psychological vulnerability, and a predisposition towards unhealthy attitudes and behaviors, may play a significant role (North and Hargreaves, 2008). Given the serious implications of censorship and the assumptions that it is predicated upon, it is necessary that researchers closely examine the relationship between listening to sexually explicit rap music and changes in attitudes, perceived norms, expectations, and behaviors. Few studies have focused on the sexual content and impact of music alone on sexual attitudes and behaviors. The research presented here addresses these issues directly.

The Influence of Media on Sexual Attitudes

The majority of research that pertains to the role of entertainment media in the sexual socialization of youth has focused on the sexual content and potential impact of primetime comedies and dramas, daytime soap operas, music videos, and magazines (North, Hargreaves, and Tarrant, 2002). An increase in the expectations of the prevalence of sexual activity, and in the acceptance of stereotypical and casual attitudes regarding sex resulting from the sustained and involved exposure to sexually oriented material, has been demonstrated in numerous studies. Greater exposure to sexual content has also occasionally been linked to a greater degree of sexual experience (e.g., Ward, 2003).

Holder-Nevins and Bain (2001) examined whether adolescent girls' sexual feelings and behavior would be influenced by exposure to sexually explicit lyrics in Jamaican dancehall music. This correlational study involved the random selection of 50 sexually active female adolescents. A second group of 50 girls who had never engaged in sexual activity was matched for age and social class. While Holder-Nevins and Bain hypothesized that Jamaican female adolescents would be more likely to engage in sexual intercourse at an early age, they instead found that this type of music was not linked to sexual feelings by either group. The questionnaire they designed to examine the hypothetical link between sexual activity and the feelings invoked by Jamaican dance hall music revealed that listening to North American soul music elicited more feelings of sexual intimacy among those who were sexually active. Although the results suggest that different genres of music have different implications for sexual attitudes and possibly for sexual behaviors, the research has some methodological limitations that must be considered when generalizing on the basis of the study's conclusions.

Holder-Nevins and Bain (2001) conducted a correlational study, which related the frequency of exposure to music to measures of various behaviors. Any effects that may result from prolonged exposure to the music in question are reflected in this particular approach; therein lies its advantage. However, the ability to attribute any effects to one specific cause, while ruling out the impact of others, is not possible. In this particular study, the correlational approach does not allow for a clear determination of whether exposure to music is the cause of certain behaviors, or vice versa. For instance: Do unsafe sexual behaviors stem from exposure to specific types of music, or is the attraction to a specific type of music borne of one's behavior (e.g., North, Hargreaves, and Tarrant, 2002)? Rather, correlational studies indicate that although there may be a relationship between rap music and certain behaviors, there may be a larger, over-arching issue that may be the true cause of both listening to this music and certain behaviors (North and Hargreaves, 2008). Although the data of Holder-Nevins and Bain show that there are indeed significant relationships, the answers to questions about causality remain unclear, and call for a between-subjects pre and post design, which would better allow for precise stimulus control and the determination of cause and effect.

The Influence of Rap Music on Sexual Aggression

In addition to pop music that contains graphic depictions of sexual acts, there exist many expressions of negative and sexist attitudes about women within rap music. Women are frequently objectified, and this has fueled concern that rap music reinforces, or justifies, the idea that coercive sexual activity is acceptable to and desired by women (Barongan and Hall, 1995). Critics argue that listening to rap music may influence people to act in more sexually aggressive and violent ways towards women (Johnson, Adams, Ashburn, and Reed, 1995). The literature on the influence of misogynistic rap music on sexual aggression against women is relevant to the broader study of the influence of sexually explicit rap in a number of ways. First, the claim that exposure to music videos containing images of women in sexually marginalized roles (of which there are similarly abundant depictions in the rap music that was addressed in this study) results in perception and attitude change, has been substantiated. In one study, participants' judgments of male-female interactions changed dramatically after watching rock videos, which contained depictions of women in sexually subordinate roles (Hansen and Hansen, 1988).

Second, a study by Johnson et al. (1995) revealed that female participants who watched rap music videos which contained images of women in sexually subordinate roles, expressed greater acceptance of teen dating violence thereafter, than did females who did not watch the videos. Johnson et al. explained these findings by citing a robust priming research literature, which indicates that exposure to verbal, visual, or auditory stimuli fosters a person's temporary access to certain pieces of information, and this in turn affects subsequent judgments and attitudes (e.g., Reyes, Thompson, and Bower, 1980; Srull and Wyer, 1980). It therefore seems plausible that exposure to sexually explicit music could potentially increase the accessibility of information related to sexual risk-taking behavior, which could then lead to greater acceptance of casual attitudes about sex, and perhaps affect the self-reported likelihood of engaging in such behaviors.

Barongan and Hall (1995) articulated a similar reasoning for the results of their study, in which a significantly greater number of men showed a sexually aggressive film vignette to a female confederate after hearing misogynistic rap music than did men who listened to neutral rap music. Barongan and Hall theorized that the misogynous rap music had a priming effect on the participants' subsequent decisions about which film to show. They argued that the medium through which this content was presented (socially acceptable music that is readily accessible to a wide population) facilitated the acceptance of sexual violence against women. The same principles were applied to the present investigation; that is, because sexually explicit rap music is not restricted in the way that pornographic materials are, the music might serve to legitimize the cognitive distortion that sexual violence is acceptable. This is important because if the listeners' condone the lyrical messages of the music, it opens up the possibility that they may be more likely to engage in those behaviors themselves.

The findings of Wester, Crown, Quatman, and Heesacker (1997) suggest that brief exposure to misogynistic rap music leads male listeners with little prior exposure to sexually violent rap music, to view their relationships with women as more adversarial, as compared with those that are not exposed to misogynistic rap music. Given these results, one might expect that men with preexisting misogynistic attitudes would manifest greater changes in attitude. Applying this thinking to the potential relationship between sexually explicit rap music, sexual attitudes, and behaviors, it seemed possible that long-term exposure to sexually explicit material might foster casual attitudes about sexual interactions by facilitating the encoding of permissive messages. The likelihood of attitude change increases when this type of encoding occurs because it enables greater attitude accessibility. Listening to sexually explicit rap may serve as a prime for individuals that are already familiar with this type of music, but for people that are unfamiliar with this type of music, sexually permissive attitudes might actually originate from short-term exposure to rap music. In other words, in addition to any longer-term effects, there might be the potential of sexually explicit rap music to momentarily influence sexual attitudes and behavior immediately after exposure.

Unpacking Media Influence: The Roles of Viewing Amount and Viewer Involvement

Collectively, the research that examines television's role in the sexual socialization process provides evidence that there is a link between the amount of time spent watching sexually oriented programming and viewer's sexual attitudes, expectations, and behaviors (see Ward and Rivadeneyra, 1999 for review). Ward and Rivadeneyra (1999) note however, that the majority of studies that examine the impact of television viewing focus solely on the *amount* of exposure to sexually oriented programming, and do not take into account other dimensions of viewing which may influence viewer's attitudes, expectations, and behaviors. In order to better understand television's role in the sexual socialization process they examined dimensions of viewing involvement. Their argument is predicated upon the idea that the degree to which viewer's feel connected to the material and are involved in the viewing experience, will dictate the extent to which they are influenced by it. This was relevant to the conceptualization of the study we present here because it is likely that factors other than amount of exposure, influence the relationship between listening to sexually explicit rap music and sexual beliefs, attitudes, and behaviors.

Ward and Rivadeneyra (1999) presented five dimensions of viewer involvement, which were based on a number of theoretical perspectives. They first considered what they termed *viewing motivation*, and defined as viewers' expected uses of television. Rubin (1994), a proponent of a uses-and-gratifications perspective, posits that people seek out and use media based on a desire to gratify specific needs. More specifically, evidence suggests that media use is more purpose driven than ritualized, thus lending support to Ward and Rivadeneyra's hypothesis that viewers who use television to obtain information about the world, would be more likely to be influenced by its content.

The second dimension of involvement, termed level of *active viewing*, addresses the ways in which differing levels of audience activity may enhance or impede media effects. Kim and Rubin (1997) argue that a person's selectivity, attention to, and involvement with the content viewed (i.e., *facilitative activity*) is a positive predictor of media effects, while avoidance, distraction, and skepticism (i.e., *inhibitory activity*) is a negative predictor of media effects. In keeping with this line of research, Ward and Rivadeneyra (1999) predicted that sexually oriented television content would have a greater impact on viewer's who watch television in a more *facilitative* manner.

The third dimension of involvement examines the *perceived realism of the portrayals*. Evidence supports the premise that television's impact will increase, the more realistic its content is perceived to be (Huston, Wartella, and Donnerstein, 1998). One individual difference that seemed likely to influence the degree to which an individual is influenced by lyrical content in the music, is how realistic that individual perceives the actions represented in the lyrical content to be. Indeed the idea that people would consider content that most closely mirrors reality more important seems like a reasonable way to process information. But while Ward and Rivadeneyra (1999) explained the role of perceived realism within the context of the cultivation theory, Taylor (2005) explained the salience of perceived realism in mediating the relationship between viewing sexual television and sexual, beliefs, attitudes, and behaviors in terms of its relationship to the social cognitive theory (SCT; Bandura, 1986, 1994, 2002) and the information-processing model (Huesmann, 1997, 1998). Bandura posits that people will be more likely to imitate a model's behavior if the models portrayal is realistic (Bandura, 2002: Bandura, Ross, and Ross, 1963).

According to the information-processing model developed by Huesmann (1997, 1998), cognitive scripts contain both procedural information (in this case, how to perform a sexual act) and declarative information (a sexual behaviors' meaning or implication), and are established through observational learning and conditioning. Scripts also impart information about when a sexual behavior is appropriate or should be expected. More specifically, the most captivating elements of a person's environment are the most likely to be attended to and incorporated into developing scripts, and this would inevitably include music, and other interesting aspects of one's environment. In this way, listening to sexually explicit rap music that describes sexual acts in realistic detail would be expected to contribute to the development of a person's sexual scripts.

The information processing model developed by Huesmann (1997, 1998) is an efficient way in which to consider the possible effects of listening to sexually explicit music, although it has primarily been utilized within the context of sexually explicit television content. Scripts guide behavior and attention to the interpretation of stimuli and media content, in this case music, and can influence and guide scripts resulting in both short-term and long-term effects. Scripts shape the development and content of ideas, which is referred to as learning, and constitutes a long-term effect, as well as influence which existing scripts are accessible for use, which constitutes a short-term effect (often referred to as priming). In short, Taylor's (2005) work suggests that listening to sexually explicit rap music could directly influence a person's behavior by guiding the expression of their beliefs and attitudes regarding appropriate sexual interaction, and that repeated exposure to sexually explicit rap music might make relevant scripts chronically available, thus informing future sexual interactions.

The fourth and fifth dimensions of involvement, *identification* and *perceived relevance* to the self, address the personal connections with television portrayals. Ward and Rivadeneyra (1999) propose that the extent to which a viewer identifies with television figures, or perceives the content to be pertinent to their own lives, will influence the degree of effect. The formulations of these dimensions are outgrowths of Greenberg's drench hypothesis (1988) and Bandura's models of Social Learning (1986; 1994). Greenberg asserts that a character portrayal, which strongly resonates with viewers may have more of an impact on attitudes, expectations, and behaviors than the sheer frequency of television characters and behaviors viewed. In a similar vein, Bandura's Social Learning Theory (1963) posits that individuals will be more likely to imitate a behavior if the model, in this case rap artists, are rewarded and not punished for their behavior.

The findings of Ward and Rivadeneyra (1999) and Taylor (2005) informed the development of the critical research questions and hypotheses, which are the basis of the study presented here. The critical research questions and hypotheses were as follows: First, we aimed to determine whether or not sexually explicit rap music influences attitudes, perceived norms, expectations, and behaviors. In keeping with the information processing model, we expected that short-term exposure to rap music with explicit sexual content would be unlikely to produce lasting changes in an individual's sexual scripts, but it might influence which behavioral scripts were used in the short term. Second, we attempted to identify a number of listener involvement characteristics that might influence the nature or degree of such effects. For the purposes of this study, listener involvement was defined to include listening motivation, level of active listening (listening habits), perceived realism, and level of identification with the performer. With regard to these four dimensions of listener involvement, we expected that the potential effects of listening to sexually explicit rap music would be stronger among viewers who were more highly motivated to listen to rap music to intentionally gather information about sex.

Third, we expected that those who were active listeners, and engaged in more facilitative activity, which includes selectivity, attention, and involvement, would be more likely to be influenced by the music because they would be more interested and invested in the content of the music. Fourth, we expected that individuals who listened to music containing graphic depictions of sexual acts and who perceived those depictions to be more realistic, would be more likely to endorse the messages espoused in the music, than individuals who listened to the same depictions and perceived them to be relatively less realistic. According to the information processing model, the belief that the lyrical content of the music is more realistic would make listeners more inclined to place importance on what they listen to, and attend to it more carefully, which would result in a greater influence of the listened music content in activating or altering existing scripts which guide behavior.

Finally, we expected that the influence of the music's sexual content would be stronger among listeners who more closely identified or connected with the portrayals. An individual may or may not strongly identify with the rappers race, gender, socioeconomic status, values, or propensity to partake in certain behaviors, and the degree to which the individual is affected by the music, will vary accordingly. Notwithstanding the fact that the relationships described above are bi-directional, and listeners' preexisting sexual attitudes and expectations affect their listening choices, the aim of the study was to examine the ways in which sexually explicit rap music's might shape listeners' conceptions of sexuality. Rap music's reach and impact will be more accurately understood when documented by systematic research on the specific variables that we report on in this chapter.

Method

Experimental Design

A between-subjects design was employed with two levels of the independent variable, one being the sexually explicit music condition (treatment), and the other being the neutral music condition (control). The factors of listening involvement were treated as independent variables in the statistical analyses. The dependent variables consisted of the participant's scores on the sexual attitudes and behavioral measures. Participants completed a series of measures, and listened to and evaluated three songs based on the condition to which they were randomly assigned.

Participants

Participants were 50 undergraduates (25 females, 25 males) recruited from an introductory psychology course at Purchase College, State University of New York. They completed a pre-test measure that included questions about their music preference, and sexual attitude measures (see Appendix A), during a regularly scheduled Introduction to Psychology class. Participants were then selected on the basis of their preference for rap music because based on previous findings, when people do not like a particular form of music, then they are not likely to pay much attention to or be influenced by it. In exchange for their participation in a "study about dating and relationships," participants received course credit or ten dollars. Participants ranged in age from 18 to 32, with a mean age of 19.4 years.

Measures of Listening Motivations and Habits

This study focused on the impact of exposure to mainstream rap music. This particular sub-genre of rap was selected above other types, because our focus was on sexual content that is heard by a wide audience and is frequently played on the radio, and to whom listeners can connect or identify. Three aspects of viewing were assessed: frequency, general involvement (5 dimensions), and perceptions of sexual content.

Frequency

To measure listening amounts, a list of was provided of 9 different musical genres. Using a 5-point scale, that ranged from 1 (*everyday / almost everyday*) to 5 (*once a month or less*), participants indicated how often they listen to music from each particular genre. Participants were also asked to indicate how many hours they listened to music on an average weekday, as well as how many hours they listened to music on a average weekend.

General Involvement

The second group of measures assessed participants' general involvement during exposure to the rap music. One measure titled "Listening Motivations" examined the degree to which listeners listen to music to learn about the world. Here participants were asked to rate each of 17 possible motivations for listening to music. Individual motives reflecting this learning and general listening motivation perspective were extracted from a measure Ward and Rivadeneyra (1999) developed. To rate each motivation, participants used a 5-point Likert scale that ranged from 1 (*strongly disagree*) to 5 (*strongly agree*). The intitial Cronbach's alpha reported for the entire scale was (α = .78), and two subscales were created to assess several aspects of listener motivation that would have particular relevance for the current research. The first subscale that was created from the composite *Listening Motivations* scale, was the *Education Learning Value* subscale which consists of 4 statements. Examples of motivations from this subscale include "because it gives me a chance to reflect on life" and "to learn about people from ethnic and cultural backgrounds different from my own." The second subscale that was created was the *Communication of Feelings* subscale, which consists of 3 statements. Examples of motivations from this subscale include "because it engenders feelings for the opposite sex" and "because I am sexually attracted to the musical performer." The final alphas were acceptable, as they exceeded .6 for each subscale (Educational α = .66, Feelings α = .72).

A second measure assessed how active or passive each participant was as a listener. This instrument was based on three types of audience activity outlined by Ward and Rivadeneyra (1999) that tap several dimensions of facilitative activity. The first aspect, *selectivity in exposure seeking*, reflects the degree to which listeners intend to enter the communication setting. The second, *involvement during exposure*, reflects the degree to which viewers either attend to the communication situation, or engage in information processing of the messages presented. The third dimension, *post-exposure use*, reflects the extent to which viewers use the messages acquired in their social or psychological life. All these dimensions were assessed in one measure titled *Music Listening Habits*. These items were based partly on items from Ward and Rivadeneyra (1999). A list of 26 statements was developed reflecting the different aspects of active listening. Examples of listening habits statements from these subscales include "I often set aside time in my day, specifically to listen to music", "I often become quite involved in the music I listen to and can easily sympathize with what the artist is singing or rapping about", and "I frequently talk to others about what music I've recently listened to." Eight items measured selectivity in exposure seeking, 17 items measured involvement during exposure, and nine items measured post-exposure use of the music. Agreement with each statement was rated using a 5-point Likert scale that ranged from 1 (*not at all like me*) to 5 (*very much like me*). Again, several individual items were removed from each subscale where results indicated that doing so would strengthen the cohesiveness of each sub-scale, and this resulted in each of the 3 subscales being reduced to 4, 11, and 8 item subscales, respectively. The final alphas exceeded .7 for each subscale (Selectivity = .70, Involvement = .73, and Use = .79).

Perceptions of Sexual Content in the Music

The remaining viewer involvement variables assessed participants' perceptions of the type of sexual content presented in the music. Participants rated the situation presented in the lyrical content, along with their level of identification with the rapper. For the identification measure, which was comprised of 8 statements about the rapper's lifestyle and actions, participants rated the extent to which they agreed or didn't agree with these statements using a 5-point Likert scale that ranged from 1 (*strongly disagree*) to 5 (*strongly disagree*). The perceived realism measure, which was comprised of 18 statements, assessed the degree to which each participant perceived the lyrical content of the music to be realistic through a measure patterned after on developed by Ward and Rivadeneyra (1999). After participants had listened to all 3 songs, they rated the songs collectively for their level of realism using a 5-point scale that ranged from 1 (*strongly disagree*) to 5 (*strongly agree*). Scores from all 3 songs were then averaged to form a composite perceived realism score.

Sexual Outcome Measures

All the measures utilized in this study, with the exception of the norm perceptions and expectations sub-scales, were adapted from measures used by Ward and Rivadeneyra (1999) to reflect the listening, rather than viewing experience. A set of sexual outcome measures were selected and/or created to reflect two dimensions of participants' sexuality: their attitudes, and their expectations and norm perceptions about their own behavior and attitudes, as well as others.

Attitudes about Sexuality

Participants' attitudes toward dating, sexual roles, and romantic relationships, were assessed using the *Attitudes toward Dating and Relationships* measure modified for the current study. Because a critical goal of this study was to determine whether exposure to sexually explicit rap increases listener's acceptance of the types of attitudes and values frequently presented in this musical genre, it was decided that the best outcome measure would be one that incorporated the exact types of values typically portrayed. Accordingly, the coding system used in Ward and Rivadeneyra's (1999) analysis of televisions sexual content was adapted to capture respondents' attitudes toward a number of themes about sexuality common in the larger culture and on television. A list of 23 statements was developed and organized into three larger categories based on this analysis. A *Men Are Sex Driven Orientation* to sexuality was represented by eight statements which defined sexual relations between men and women in terms of men's inability to resist sexual urges and remain monogamous. Here, sex is seen as the natural outgrowth of men's readiness and willingness to have sex. Six statements, which portrayed dating, and sexual relations as a game or competition represented a *Recreational Orientation* to sexuality. Here, the sexual double standard prevails, and males and females are viewed as being on opposing teams ("battle-of-the-sexes"), taking different roles, and using sex as the object of exchange. Finally, a *Women Are Sexual Objects Orientation* to sexuality was represented by nine statements, which emphasized that women should expect their interactions with men to be solely based on their physical appearance. Participants rated their level of agreement with each statement using a 5-point Likert scale that ranged from 1 (*strongly disagree*) to 5 (*strongly agree*). A full formatted version of the sexual attitudes measure (pre and post) is included in Appendix B.

Prior to any subsequent statistical analyses, reliability analyses of this measure and its three subscales were conducted. The *Men Are Sex Driven Orientation* and *Women Are Sexual Objects Orientation* subscales were reliable at both times (time 1 $\alpha = .77$, time 2 $\alpha = .82$) and (time 1 $\alpha = .67$, time 2 $\alpha = .66$), respectively. Examples of statements from these subscales include "It's difficult for men to resist sexual urges and to remain monogamous", "An attractive woman should expect sexual advances and learn how to handle them", and "A good way to reward or punish someone is by giving or withholding sex." Two items were removed from the *Recreational Orientation* subscale to improve the cohesiveness of this subscale, bringing the final alphas to .65 (time 1) and .62 (time 2).

Sexual Expectations and Perceived Norms

To assess participants' expectations and perceived norms about the level of sexual activity among their peers and condom use, 4 out of the 12 risk-related or recreational sexual behaviors that Ward had selected from the general literature on sexual risks (e. g., Metzler Noell, and Biglan, 1992) were revised using the wording from the information, motivation, and behavioral skills (IMB) model of AIDS risk behavior change (J.D. Fisher and Fisher, 1992; W. A. Fisher and Fisher, 1993). The other 7 statements, which pertained to condom use, were drawn directly from this model. For each statement, participants were asked to rate their level of agreement with each statement using a 5-point Likert scale that ranged from 1 (*strongly disagree*) to 5 (*strongly agree*). The four items that pertained to recreational sexual behaviors, and 2 items that pertained to condoms use were removed because results indicated that doing so would strengthen the cohesiveness of the scale, and led to the formation of the *Perceived Social Normative Support for Condom Use* subscale, which was reliable at both times (time 1 $\alpha = .84$, time 2 $\alpha = .89$). Examples of the statements include "Most people that are important to me think I should buy latex condoms" and "Most people that are important to me think my partner(s) and I should always use latex condoms during sexual intercourse." A complete list of items that comprise the scales and subscales for all of the listener involvement and sexual outcome measures, as well as their reliabilities, is presented in Appendix C.

Sexual Behavior

In addition to measuring the effects of rap music on sexual attitudes, the experimenter wanted to explore whether or not there could be a positive effect of this music on behavior. One of the most important aspects of safe sex is the use of condoms, so a condom-related behavioral measure was included, which was to see how many condoms each participant would take after listening to the music.

Stimuli

To examine the potential effects of sexually explicit rap music on sexual attitudes and behaviors, a number of rap artists were identified that were known both from research and personal knowledge, to write, record, and produce songs that are highly sexual in nature. From a play-list of roughly 35 songs, the song-list was narrowed down to 6 songs: 3 that are sexually explicit in nature, and 3 that are not. The songs that were selected were released after 1995 because it was important that they be popular and commercial, and reach a broad audience. In essence, the goal was to investigate music that is played often in various media formats, and has the potential to be all the more influential because of its wide distribution. To play music during experimental sessions, a Sony 5-Disc CD changer was connected to a 4-Channel PreSonus headphone amplifier. Participants listened to the music through Sony MDR-7505 Headphones.

Participants in the sexually explicit music condition listened to three sexually explicit rap songs. In the treatment condition, the three songs were: (a) "Wait (The Whisper Song)" (Ying Yang Twins, 2005, track 10); (b) "Play" (David Banner, 2005, track 5); and (c) "Oochie Wally" (QB Finest, 2000, track 7). The rap songs in this group contained frequent and explicit references to sex. Participants in the neutral music condition also listened to three non-sexually explicit rap songs. In the control condition the five songs were: (a) "Got Yourself a Gun" (Nas, 2001, track 3); (b) "Shook Ones, Pt.II" (Mobb Deep, 1995, track 15); and (c) "Get By" (Talib Kweli, 2002, track 3). These songs didn't contain any significant references to sex and were primarily concerned with partying, making money, and references to the problems of social injustice facing African-Americans in America.

Procedure

As mentioned earlier, participants completed a pre-screen measure that included questions about their music preference, and the pre-sexual attitude measures, during a regularly scheduled Introduction to Psychology class. They were selected to participate in the study based on their preference for rap music, and each participant was tested individually so that they would not be influenced by other participants' reactions to the music. As each of the participants arrived for his/her individual sessions, the experimenter asked the participant to take a seat at the desk adjacent to the experimenter experimenter's desk. Then the experimenter explained that they would be participating in a study to determine their attitudes towards a number of social topics, in particular, listeners' perceptions of dating and relationships as they are portrayed in rap music. Participants were assured of their anonymity and given an opportunity to withdraw from the experiment at any point without penalty. All participants agreed to participate and provided informed consent.

Following completion of the informed consent pages, participants were handed the first experimental packet and instructed to complete a series of questions about demographic and music preference information, followed by a number of measures in which they were asked to rate how strongly they agreed or disagreed with a number of statements pertaining to listening motivations and habits using a 5-point Likert scale. Upon completion of these pages, the experimenter handed the participant the second experimental packet, and began to play the first of three songs for the participants according to the experimental condition to which they were randomly assigned.

Participants in both conditions were asked not to begin writing their responses to the music until playback was completed. After each song played, participants completed a series of questions about the theme of the music to determine their comprehension of the lyrics, as well as indicate how discomforting the music was, and whether or not they had listened to the song prior to this experiment. Participants also answered questions about how sexually explicit they perceived the music to be, and whether or not they liked or disliked the song. In addition, they were asked to rate their level of identification with the rapper using a 5-point Likert scale. The experimenter observed the progress of participants as they completed the items. Once the participant had completed all of the items, the next song was played and the sequence was initiated again.

At this point, the participants were informed that the experimenter needed to step outside into the hall for a moment, and were handed a third experimental packet which they could complete by themselves. The experimenter wanted to give participants a chance to complete the measures privately, and also create the opportunity for them to take condoms, a sexually related behavior, which could be measured in a fairly unobtrusive way. In addition to a bowl of condoms, there were two bowls of candy placed on the desk in order to ensure that the experimental setting wasn't awkward and mimicked typical college health settings in which students can obtain condoms. At this point, the participants were asked to rate the level of perceived realism collectively for all songs using a 5-point Likert scale, after which they completed the post-sexual attitudes measures, which were identical to those administered in the pre-screen measure.

After the participant exited the experimental room with packet in hand, the experimenter asked the participant what they thought the study was about (manipulation check), after which they were provided with a copy of the debriefing. To ensure the integrity of the participant pool, participants were instructed to avoid any discussion of experimental details until after all experimental sessions were completed. Following the participant's departure, the experimenter reentered the experimental room to count and record the number of condoms present in the bowl. For the typical participant, the duration of the study was 45 minutes.

Results

Preliminary Analysis

Descriptive statistics of the independent and dependent variables are provided in Table 1 for each condition. In general, there was little variation in the way people responded to questions regarding their listening behaviors and sexual attitudes. Preliminary analyses were also conducted examining zero-order correlations between the key viewer involvement variables and the sexual outcome measures.

Table 1. Descriptives of Key Variables by Condition

	Treatment Group		Control Group	
Variable	**Mean**	**S.D.**	**Mean**	**S.D.**
General Reactions				
Average Liking Across Songs	2.41	1.03	3.75	.89
Sexual Explicitness of Songs	4.69	.63	1.68	.60
Discomfort Level	2.83	1.35	1.37	.53
Viewer Involvement				
Listening Motivation (overall)	3.42	.44	3.37	.57
Educational/Learning Value	2.99	.73	2.94	.88
Communication of Feelings	2.81	.92	2.63	.82
Listening Habits (Active Listening)				
Selectivity Before	3.13	.83	3.28	1.00
Involvement During	3.22	.44	3.39	.72
Use After	2.92	.61	3.15	.87
Identification	1.61	.66	2.43	.75
Perceived Realism	2.60	.50	2.89	.61
Sexual Outcome Measures				
Men Are Sex Driven	-.20	.49	-.09	.62
Dating Is A Game	.01	.45	.12	.67
Women Are Sexual Objects	-.19	.50	-.05	.46
Perceived Social Norm Support For Condom Use	.14	.69	13	.82
# of Condoms Taken	2.84	3.56	1.32	2.16

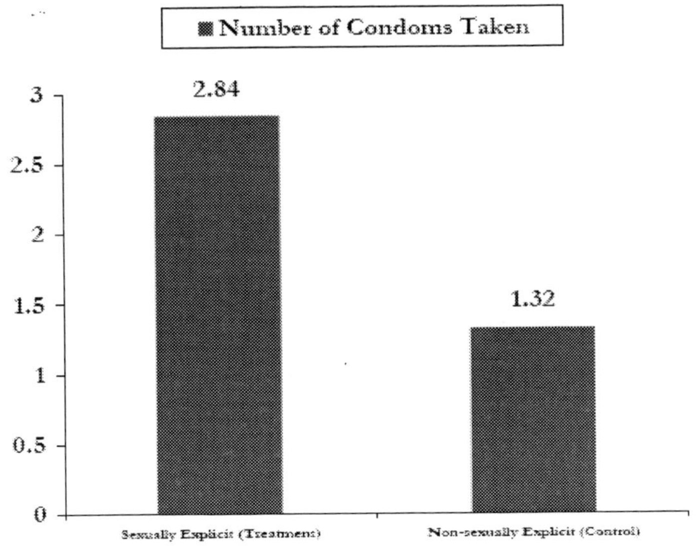

Note. Marginally significant difference ($F(1,48) = 3.3$, $p = .07$) in the number of condoms taken based on experimental condition.

Figure 1. Number of condoms taken based on experimental condition.

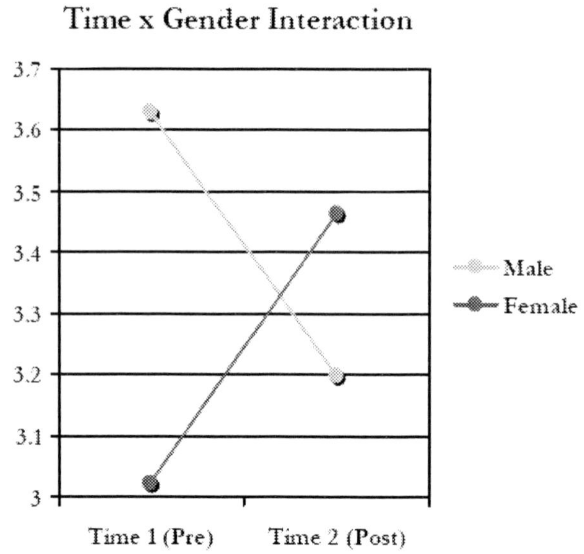

Note. Significant interaction of time and gender on perceived social normative support for condoms within the treatment condition ($F_{(1, 21)} = 9.91, p < .01$).

Figure 2. Perceived norm support for condom use at time 1 and time 2.

Overview of Analysis

For each of the outcome variables of interest, a repeated measures analysis of variance (ANOVA) was conducted to compare differential rates of change between participants in control and treatment conditions. This analytic strategy was employed because it accounts for change in participants' scores over time, while testing whether other factors, including the five dimensions of viewer involvement, had a significant effect on the change of scores from time 1 to time 2. By computing these change scores and by conducting repeated measures analyses that examine differences over time, the change that occurs within participants is relatively isolated with respect to other variables. Analyses are based on differential rates of change, which minimizes the impact of pretest differences, thus limiting the potential effects of factors that are not of direct concern to the researcher.

The first set of analyses tested the premises of the information processing model. More specifically, whether listening to rap music would have an affect on sexual attitudes, expectations, and perceived norms. We hypothesized that short-term exposure to this type of music would not produce lasting changes in attitudes expressed thereafter. In order to examine this, a series of 2 (condition) x 2 (time) mixed design ANOVAs were conducted, with condition as the between subjects variable, and time as the within subjects variable. Results of each ANOVA indicated no significant main effects of condition or time, and no significant interactions on sexual attitudes, expectations, or perceived norms (all F's < 1).

To test the hypothesis that the music would influence short-term behavior, a independent samples t-test was conducted on the number of condoms taken. Results indicated that there was a marginally significant difference between conditions, such that participants in the sexually explicit condition (M = 2.84, SD = 3.55) took more condoms that those in the non-sexually explicit condition (M = 1.32, SD = 2.16), t (48) = 31.83, p = .07. The large range (0-

12 condoms taken) coupled with the size of the standard deviation accounts for the marginal significance (Refer to Figure 1).

While the there was no significant interaction between condition and pre and post sexual attitudes measures, there were a few noteworthy relationships that emerged when we treated gender as an independent variable. Initially, change scores were computed (time 2-time 1), and differences in change were examined via a linear regression analysis. This indicated that there was change over time in sexual attitudes, but that the change was localized among gender within the treatment condition. Two 2 (condition) x 2 (gender) x 2 (time) mixed ANOVAs were conducted in which time was treated as the within subjects variable, with the perceived social norm support for condom use and women as sexual objects subscales as the dependent variables. Within the treatment condition, there was a significant time by gender interaction, indicating that women became more likely to have the perception that the social norm approves or recommends the use of condoms, while men became less likely to affirm or to have the perception that the social norm approves or recommends the use of condoms ($F_{(1, 21)} = 9.91$, $p < .05$; Refer to Figure 2). In addition, there was a significant time by gender interaction for the women are sexual objects subscale, indicating that women changed in the direction of being less likely to affirm that women should think about or expect their interactions with men to be based on sexuality and appearance, while men were more likely to affirm that women should expect their interactions with men to be based on sexuality and appearance ($F_{(1, 19)} = 4.39$, $p < .05$). In the control conditions, no such interactions were observed, the relationships holds when controlling for listening motivation (Refer to Figure 3).

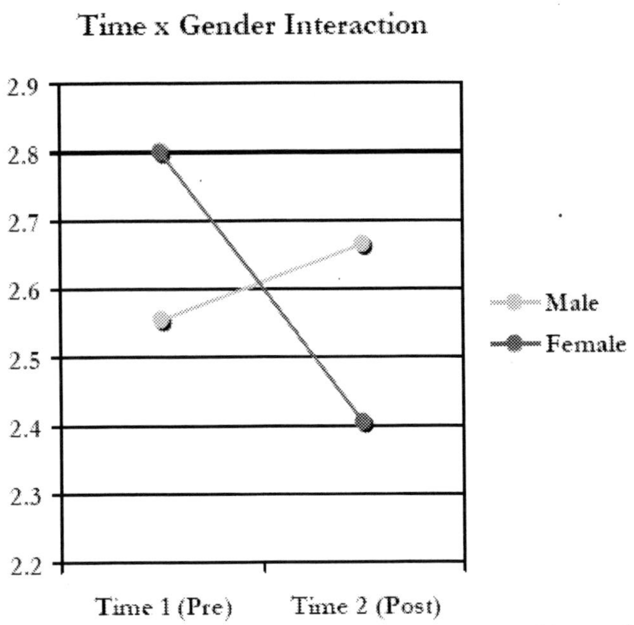

Note. Significant interaction of time and gender on women are sexual objects subscale ($F_{(1, 19)} = 4.39$, $p < .05$).

Figure 3. Attitudes about viewing women as sexual objects at time 1 to time 2.

We also predicted that a number of dimensions of listening involvement would influence the extent to which participants were affected by the music. This hypothesis was tested by a MANCOVA in which condition and gender were treated as independent variables, and the five dimensions of viewing involvement were included as covariates in the analysis. The change scores for sexual attitudes, perceived norms, and expectations were included as the dependent variables. Contrary to our initial hypothesis, the five dimensions of viewing involvement, which were treated as covariates in the analysis, did not influence the extent to which the manipulated variable affected the outcome measures. In addition, there were no significant three-way interactions between the treatment, viewer involvement variables, and change in pre and post sexual attitude measures.

Correlational analyses were conducted, between the five viewer involvement variables and the sexual outcome variables. These results are summarized and presented in Table 2. For the most part, the viewer involvement variables did not correlate with the participants' sexual attitudes, expectations, perceived norms, or behavior. However, several interesting correlations emerged between gender, condition, viewer involvement variables, sexual outcomes, and general reactions to the music, which speak to the power of the manipulation (refer to Table 3). First it's important to note that there was a negative correlation between condition and the sexual explicitness of the song ($r = -.93$, $p < .01$), which indicates that people in the treatment condition found the music to be sexually explicit, and were discomforted by it ($r = -.59$, $p < .01$). In addition, there was a positive correlation between condition and average liking across the songs ($r = .58$, $p < .01$), such that people in the control condition liked the music more than those in the treatment condition (refer to Table 4). Participants also identified more with the rappers in the control condition ($r = .51$, $p < .05$), than those in the treatment condition. With respect to gender, there was an unexpected positive correlation between gender and communication of feelings ($r = .29$, $p < .05$), making it appear that women more than men, listen to music in order to express or communicate their internal feelings to another person, or because it engenders feelings for the opposite sex.

Table 2. Correlations between Viewer Involvement Variables and Sexual Outcomes

	Men Are Sex Driven Change from Time1 to Time 2:	**Dating Is A Game**	**Women Are Sexual Objects**	**Perceived Social Norm Support For Condoms**	**# of Condoms Taken**
Viewer Involvement					
Listening Motivation (overall)				.32*	.28*
Educational/Learning Value					
Communication of Feelings				.32*	.30*
Listening Habits (Active Listening)					
Selectivity Before			.35*		
Involvement During					
Use After				.30*	
Identification					
Perceived Realism					
Active Viewing					

* $p < .05$. ** $p < .01$. *** $p < .001$.

Table 3. Correlations between Gender/Condition and Viewer Involvement Variables/Sexual Outcomes/General Reactions

	Gender	Condition
General Reactions		
Average Liking Across Songs		.58**
Sexual Explicitness of Songs		-.93**
Discomfort Level		-.59**
Viewer Involvement		
Listening Motivation (overall)	.36*	
Educational/Learning Value		
Communication of Feelings	.29*	
Listening Habits (Active Listening)		
Selectivity Before		
Involvement During		
Use After		
Identification		.51*
Perceived Realism		
Sexual Outcome Measures		
Men Are Sex Driven		
Dating Is A Game		
Women Are Sexual Objects		
Perceived Social Norm Support	.44*	
For Condom Use		
# of Condoms Taken		

* $p < .05$. ** $p < .01$. *** $p < .001$.

In summary, with respect to the manipulation, there was no main effect of condition: the experimental manipulation did not influence change between the pre and post sexual attitude measures. The extent to which the manipulated variable affected the outcome measures was not influenced by the viewer involvement variables. There were also no significant three-way interactions between the treatment, viewer involvement variables, and change in pre and post sexual attitude measures. However, in the treatment condition we found that there was a significant time by gender interaction in the perceived norms about condom use subscale, indicating that women became more likely to have the perception that the social norm approves or recommends the use of condoms, while men became less likely to affirm or to have the perception that the social norm approves or recommends the use of condoms. There was also a significant time by gender interaction in the treat condition with regard to the women are sexual objects subscale, which indicates that women changed in the direction of being less likely to affirm that women should think about, or expect, their interactions with men to be based sexuality and appearance, while men were more likely to affirm that women should expect their interactions with me to be based on sexuality and appearance. In terms of the behavioral measure, there was a marginally significant difference in the number of condoms taken based on the experimental condition, such that participants in the treatment condition took more condoms than those in the control condition.

Table 4. Correlations between General Reaction Measures and Viewer Involvement Variables

	Liking Average	Sexual Explicitness	Discomfort Level	Listening Motivation	Learning Value	Comm. Feelings	Habits Before	Habits During	Habits After	Identification	Perceived Realism
General Reactions		-.62**	-.66**								
Average Liking Across Songs											
Sexual Explicitness of Songs											
Discomfort Level											
Viewer Involvement Variables											
Listening Motivation (overall)				.81**							
Educational/Learning Value						.49**	.37**	.49*	.58**		30*
Communication of Feelings				.73**			.32*		.32*		
Listening Habits											
Selectivity Before				.39**							
Involvement During				.47**			.43**		.76**		
Use After				.53**			.63**				.35*
Identification	.71**	-.53**	-.58**								.58**
Perceived Realism	.61**	-.31*	-.37*								

* $p < .05$. ** $p < .01$. *** $p < .001$.

Conclusion

The goal of this study was to examine how music listening might relate to young adults' sexual attitudes and behaviors, using an experimental research design. The pattern of results indicates that in general, there was no significant relationship between listening to sexually explicit rap music and sexual attitudes, perceived norms, and expectations. There may be some differences based on gender, but we can't say definitively that it was related to exposure to sexually explicit rap music because of the lack of a significant three-way interaction.

In addition, the extent to which the music affected the outcome measures was not influenced by the viewer involvement variables that were examined. A marginally significant effect was found with regards to participants' behavior, such that those assigned to the sexually explicit condition took a marginally greater number of condoms than those in the control condition. Thus, it appears as if listening to sexually explicit rap music does not influence sexual attitudes per se, but may influence behavior in the short term. This essentially lends support to the in the information processing model, which posits that exposure to various forms of media will be unlikely to produce lasting changes in an individual's sexual scripts, but may influence which scripts are used to guide decisions and behaviors in the short term. In this case, short-term exposure to sexually explicit rap did not produce lasting changes in attitudes expressed thereafter, but it did serve as a prime, which influenced which scripts were used to guide behavioral decisions in the short term. This conceptual model accounts for the lack of results in regards to sexual attitudes, and at the same time explains the marginal significance in the number of condoms taken.

These findings should be interpreted cautiously, however, for a number of reasons. While this experimental research allowed us to directly investigate whether exposure to sexually explicit rap music caused sexual attitudes, perceived norms, expectations, and behaviors to change, it was also prone to demand characteristics, such that the participants might have been able to guess the hypothesis under investigation, which would guide their response to the stimuli. It is reasonable to expect that the majority of our participants were aware of the link between exposure to sexually explicit rap and sexual attitudes and behaviors. Accordingly, if they are asked to listen to sexually explicit rap music, and then answer a number of questions about their sexual attitudes, it is likely that this conscious awareness of what is 'expected' could plausibly influence their responses. Given that they were psychology undergraduates, they were probably the most highly sensitized group of people to the hypothetical effects of rap music on behavior.

Second, it is not clear whether the effects of rap music found within an experimental setting translate to real-world environments. Even if this research provided unequivocal support for a rap music-sexual permissive attitude link, we could not say with confidence that listening to this music precipitates sexual risk-taking behaviors, just merely that exposure to this music could predispose a person toward engaging in that kind of behavior. It also seems that a person's recognition of the real consequences of sexual risk taking behavior would diminish any sexual proclivities primed by the music (North and Hargreaves, 2008).

Third, in focusing on the lyrical content of rap music, it is natural to assume that the marginally significant behavioral effect in this study is attributable solely to the lyrical content, but there is also the actual music to consider. This study raises interesting questions about the power of music alone, and the potential interplay between the music, lyrics, and attitudes and behaviors. Perhaps it is wise to think about this behavioral effect as being not only influenced by scripts but also by specific elements of the music (i.e., rhythm, harmony, melody, structure, texture, etc.) and the overall emotional quality of the music.

It is also important to consider that the context in which they listened to this music may not have been conducive to attitude and norm perception change. Many of the participants indicated that they listen to this type only within nightclubs or social gatherings. Interestingly, many women noted that they listen to this kind of rap solely for the beat and music and do not in any way condone the sexual content of the lyrics. This suggests that people may be deliberately ignoring or overlooking the lyrical content of the music in order to enjoy the rhythmic and melodic component of the music, which many participants noted "was the best kind of music to dance to." In addition, many participants indicated that they liked rap music generally but that they listened to other types of rap music that did not contain explicit sexual content. Just because participants indicated that they liked rap on the pre-screen measures doesn't necessarily mean that they listen to or enjoy this particular sub-type of rap.

The lack of direct experimental results could also be attributed to the fact that the listeners were college undergraduates, most of whom are sexually experienced. It is possible that this population of college students would be less susceptible to absorbing messages espoused in the music by virtue of their older age and greater level of sexual experience. But what about younger people? Different effects might be observed among younger audiences whose beliefs and attitudes about sex might not yet be established (Taylor, 2005).

Indeed, this study and Ward's (2003) raise a number of interesting questions that could inform further research concerning the impact of sexually explicit rap music on listeners. First, do listeners view this music in a positive or negative way, and even if it's in neutral way, what are the implications of this passive acceptance, however benevolent? Is this simply a reflection of the way younger generations view sex in an increasingly uninhibited media culture? In other words, we need to gather more information about how listeners interpret the messages in rap music. Second, we also need to begin to investigate how reactions to this music are influenced by participants' previous music listening history and cultural and ethnic background. By extension, researchers should expand the demographics of samples tested beyond North America because different countries have vastly different cultural attitudes about sex in general, along with its role and representation in the arts (North and Hargreaves, 2008). Third, future research could directly investigate the role that intervening variables such as self-esteem, family environment, education, and socio-economic status play in the relationship between listening to sexually explicit rap music and sexual attitudes and behaviors. In short, what seems clear from the limited evidence available as well as the new data that we present here is that the relationship between sexually explicit rap music and sexual attitudes and behaviors is quite complex and warrants additional consideration.

Appendix A. Full-Formatted Version of Pre-screen Measure

Pre-screen to Participate in a Study about Music and Social Attidues

Purchase College, State University of New York

Purpose: You are being asked to complete a brief questionnaire in order to determine your eligibility in this study based on a number of criteria. If you meet the criteria for participation this study, you may be contacted by the experimenter, and invited to participate in the main study either for course credit or $10.

Principal Investigator: Lucrezia Alcorn

Co-Investigator: Dr. Anthony Lemieux

Confidentiality: The background information collected in the form of this questionnaire will be kept confidential at all times. If you are invited and agree to participate, your name will not appear on any data collection sheet or questionnaire. Instead, you will be assigned a numerical participant code which is used in assigning questionnaires and tracking results. Only the researchers will see your name and subject number. All data you provide will be kept in the strictest confidence.

Your signature acknowledges that you have read the information and willingly sign this consent form.

I have read the above statements and I understand my rights.

____

Name (Signature)

••

Please provide the following information in the case that you are contacted to participate in the main study.

____

Name (Printed)

____

Phone Number

____

Email Address

••

Please tell us how much you like each of the following types of music using the scale shown below. Write your response on the line to the left of each style of music.

1 = Dislike Very Much 2 = Dislike A Little 3 = No Opinion 4 = Like A Little 5 = Like Very Much

_____RandB
_____Country
_____Pop
_____Classical
_____Rock
_____Jazz
_____Reggae
_____Rap/Hip-Hop

Attitudes about Dating and Relationships

There are lots of opinions about dating and relationships. We are interested in what you think. Below are statements that reflect different opinions. Please read each statement and decide how much you agree or disagree with it. Then for each statement, check the *one* response option that best reflects your personal opinions.

SD = Strongly Disagree D = Disagree N = Neither Agree Nor Disagree A = Agree SA = Strongly Agree

1. Men are mostly interested in women as potential sex partners and don't want to be "just friends" with a woman.
2. It's difficult for men to resist sexual urges and to remain monogamous.
3. It is natural for a man to want to admire or ogle women and to comment on their bodies, even if he has a girlfriend.
4. Something is wrong with a guy who turns down a chance to have sex.
5. Men who are "good with the ladies" and who can get any woman into bed are cool.
6. Men are always ready and willing for sex; they think about it all the time.
7. It is only natural for a man to make sexual advances toward a woman he finds attractive.
8. Men are sex driven and have trouble being faithful.
9. Dating is basically a game, a battle of the sexes, where both males and females try to gain the upper hand and manipulate each other.
10. Dating is a game or recreational sport.
11. Sexual activity is desirable as early in a relationship as possible.
12. A good way to reward or punish someone is by giving or withholding sex.
13. You don't need a relationship to have good sex; all you need are two people who are attracted to each other.
14. A man will be most successful in meeting or picking up women if he has a "rap" or uses flattering, sexy, or cute pick-up lines.
15. Women are sexual objects whose value is based on their physical appearance.
16. An attractive woman should expect sexual advances and should learn how to handle them.
17. Women should be more concerned about their appearance than men.
18. Using her body and looks is the best way for a woman to attract a man.
19. Women should spend a lot of time trying to be pretty; no one wants to date a woman who has "let herself go."
20. There's nothing wrong with men whistling at shapely women.
21. It bothers me when a man is interested in a woman only if she is pretty.
22. There is nothing wrong with men being primarily interested in a woman's body.
23. Being with an attractive woman gives a man prestige.
24. Most people that are important to me have had sexual intercourse at least once.
25. Most people that are important to me have had sex on a hook-up or with someone that they just met.
26. Most people that are important to me have gotten drunk and engaged in sexual behavior that they later regretted.
27. Most people that are important to me have intentionally use alcohol or drugs to lose inhibitions about sex.
28. Most people that are important to me have had more than 10 sexual partners overall.

29. Most people that are important to me will usually have sex without using condoms.
30. Most people that are important to me think I should talk about safer sex with my partner(s) before having sex with them.
31. Most people that are important to me think I should try to persuade my partner(s) to practice only safer sex.
32. Most people that are important to me think I should buy latex condoms.
33. Most people that are important to me think I should always have latex condoms handy.
34. Most people that are important to me think my partner(s) and I should always use latex condoms during sexual intercourse.

Appendix B. Full-Formatted Version of Sexual Attitudes Measure

Attitudes about Dating and Relationships

There are lots of opinions about dating and relationships. We are interested in what you think. Below are statements that reflect different opinions. Please read each statement and decide how much you agree or disagree with it. Then for each statement, check the *one* response option that best reflects your personal opinions.

SD = Strongly Disagree D = Disagree N = Neither Agree Nor Disagree A = Agree SA = Strongly Agree

1. Men are mostly interested in women as potential sex partners and don't want to be "just friends" with a woman.
2. It's difficult for men to resist sexual urges and to remain monogamous.
3. It is natural for a man to want to admire or ogle women and to comment on their bodies, even if he has a girlfriend.
4. Something is wrong with a guy who turns down a chance to have sex.
5. Men who are "good with the ladies" and who can get any woman into bed are cool.
6. Men are always ready and willing for sex; they think about it all the time.
7. It is only natural for a man to make sexual advances toward a woman he finds attractive.
8. Men are sex driven and have trouble being faithful.
9. Dating is basically a game, a battle of the sexes, where both males and females try to gain the upper hand and manipulate each other.
10. Dating is a game or recreational sport.
11. Sexual activity is desirable as early in a relationship as possible.
12. A good way to reward or punish someone is by giving or withholding sex.
13. You don't need a relationship to have good sex; all you need are two people who are attracted to each other.
14. A man will be most successful in meeting or picking up women if he has a "rap" or uses flattering, sexy, or cute pick-up lines.
15. Women are sexual objects whose value is based on their physical appearance.
16. An attractive woman should expect sexual advances and should learn how to handle them.
17. Women should be more concerned about their appearance than men.
18. Using her body and looks is the best way for a woman to attract a man.

Appendix B. Continued.

19. Women should spend a lot of time trying to be pretty; no one wants to date a woman who has "let herself go."
20. There's nothing wrong with men whistling at shapely women.
21. It bothers me when a man is interested in a woman only if she is pretty.
22. There is nothing wrong with men being primarily interested in a woman's body.
23. Being with an attractive woman gives a man prestige.
24. Most people that are important to me have had sexual intercourse at least once.
25. Most people that are important to me have had sex on a hook-up or with someone that they just met.
26. Most people that are important to me have gotten drunk and engaged in sexual behavior that they later regretted.
27. Most people that are important to me have intentionally use alcohol or drugs to lose inhibitions about sex.
28. Most people that are important to me have had more than 10 sexual partners overall.
29. Most people that are important to me will usually have sex without using condoms.
30. Most people that are important to me think I should talk about safer sex with my partner(s) before having sex with them.
31. Most people that are important to me think I should try to persuade my partner(s) to practice only safer sex.
32. Most people that are important to me think I should buy latex condoms.
33. Most people that are important to me think I should always have latex condoms handy.
34. Most people that are important to me think my partner(s) and I should always use latex condoms during sexual intercourse.

Appendix C. Items That Make up Each sub Scale and Associated Reliabilities

Scale and Sub-Scale Title	Item (* indicates reverse coding)	Reliability at Time 1 (α)	Reliability at Time 2 (α)
Sexual Attitudes: Men Are Sex Driven (subscale)	1.) Men are mostly interested in women as potential sex partners and don't want to be "just friends" with a woman.	.77	.82
	2.) It's difficult for men to resist sexual urges and to remain monogamous.		
	3.) It is natural for a man to want to admire or ogle women and to comment on their bodies, even if he has a girlfriend.		
	4.) Something is wrong with a guy who turns down a chance to have sex.		

Scale and Sub-Scale Title	Item (* indicates reverse coding)	Reliability at Time 1 (α)	Reliability at Time 2 (α)
	5.) Men who are "good with the ladies" and who can get any woman into bed are cool.		
	6.) Men are always ready and willing for sex; they think about it all the time.		
	7.) It is only natural for a man to make sexual advances toward a woman he finds attractive.		
	8.) Men are sex driven and have trouble being faithful.		
Sexual Attitudes: Dating Is A Game (subscale)	1.) Dating is basically a game, a battle of the sexes, where both males and females try to gain the upper hand and manipulate each other.	.65	.62
	2.) Dating is a game or recreational sport.		
	3.) Sexual activity is desirable as early in a relationship as possible.		
	4.) A good way to reward or punish someone is by giving or withholding sex.		
Sexual Attitudes: Women Are Sexual Objects (subscale)	1.) Women are sexual objects whose value is based on their physical appearance.	.67	.66
	2.) An attractive woman should expect sexual advance and should learn how to handle them.		
	3.) Women should be more concerned about their appearance than men.		
	4.) Using her body and looks is the best way for a woman to attract a man.		
	5.) Women should spend a lot of time trying to be pretty; no one wants to date a woman who has "let herself go."		
	6.) There's nothing wrong with men whistling at shapely women.		
	7.) It bothers me when a man is interested in a woman only if she is pretty.*		
	8.) There's nothing wrong with men being primarily interested in a woman's body.		
	9.) Being with an attractive woman gives a man prestige.		

Appendix C. Continued

Scale and Sub-Scale Title	Item (* indicates reverse coding)	Reliability at Time 1 (α)	Reliability at Time 2 (α)
Sexual Attitudes: Perceived Social Norm Support for Condoms (subscale)	1.) Most people that are important to me think I should talk about safer sex with my partner(s) before having sex with them.	.84	.89
	2.) Most people that are important to me think I should try to persuade my partner(s) to practice only safer sex.		
	3.) Most people that are important to me think I should buy latex condoms.		
	4.) Most people that are important to me think I should always have latex condoms handy.		
	5.) Most people that are important to me think my partner(s) and I should always use latex condoms during sexual intercourse.		
Listening Habits: Selectivity Before Exposure (subscale)	1.) I usually get my favorite artists albums as soon as they come out.	.70	
	2.) I often set aside time in my day, specifically to listen to music.		
	3.) There are certain musical artists that I really make a point to follow all of their music.		
	4.) I usually attempt to listen to an artist's album from start to finish.		
Listening Habits: Involvement During Exposure (subscale)	1.) While listening to music, I often try to track the lyrical/musical development of an artist from one song/album to another.	.73	
	2.) While listening to music, I often try to pick out the flaws and mistakes.		
	3.) While listening to music, I often try to pick out the subtle intricacies in the music.		
	4.) I generally get frustrated if I have trouble following the lyrics, or if the lyrics don't make any sense to me.		
	5.) While listening to music, I often notice and comment on the quality of the music or the aesthetic presentation of the artist.		
	6.) I often become quite involved in the music I listen to and can easily sympathize with what the artist is singing about.		
	7.) I frequently sing along to the music.		
	8.) While listening to the music, I often think about how the situations presented in the lyrics relate to situations in my own life.		

Scale and Sub-Scale Title	Item (* indicates reverse coding)	Reliability at Time 1 (α)	Reliability at Time 2 (α)
	9.) While listening to a music album, I often think about how the theme of the music develops across the entire album.		
	10.) I often try to guess how a particular musical artist that I am not familiar with, will sound.		
	11.) While listening to certain types of music, I often make judgments about the appropriateness of the morals and messages presented.		
Listening Habits: Use After Exposure (subscale)	1.) Sometimes people described in the lyrics of songs are in my dreams.	.79	
	2.) Sometimes the musical artists I listen to are in my dreams.		
	3.) I often think about how the people described in the lyrics of the music are similar to people I know.		
	4.) Frequently lyrical expressions, musical styles, or particular themes or situations presented in music will spark interesting discussions.		
	5.) I often think about the particular of music I listen to a few days after I've listened to it.		
	6.) I frequently talk to others about what music I've recently listened to.		
	7.) I often have given people advice based on messages I've gathered from lyrical messages presented in the music I listen to.		
	8) I often retell stories or talk about information about the music and musical artists I listen to.		
Listening Motivations: Educational / Learning Value (subscale)	1.) I listen to music: to learn about people from ethnic and cultural backgrounds different from my own.	.66	
	2.) I listen to music: because it teaches me things not learned in school.		
	3.) I listen to music: because I look up to musical artists as role models.		
	4.) I listen to music: because it gives me a chance to reflect on my own life.		
Listening Motivation: Communi-cation of Feelings (subscale)	1.) I listen to music: in order to express and communicate my internal feelings to another person.	.72	
	2.) I listen to music: because it engenders feelings for the opposite sex.		
	3.) I listen to music: because I am sexually attracted to the musical performer.		

References

Bandura, A. (1986). *Social foundations of thought and action: A social cognitive theory.* Englewood Cliffs, NJ: Prentice Hall.

Bandura, A. (1994). *Social cognitive theory of mass communication.* In J. Bryant and D. Zillmann (Eds.), Media effects: Advances in theory and research (2nd Edition) (pp. 60-90). Hillsdale, NJ: Lawrence Erlbaum Associates.

Bandura, A. (2002). Social cognitive theory of mass communication. In J. Bryant and D. Zillmann (Eds.), *Media effects: Advances in theory and research* (2nd Edition) (pp. 121-154). Mahwah, NJ: Lawrence Erlbaum Associates.

Bandura, A., Ross, D., and Ross, S.A. (1963). Imitation of film-mediated aggressive models. *Journal of Abnormal and Social Psychology*, **66**, 3-11.

Banner, David. (2005). Play. On Certified [CD]. Santa Monica, CA: Universal Music Group.

Barongan, C., and Hall, G.C.N. (1995). The influence of misogynous rap music on sexual aggression against women. *Psychology of Women Quarterly*, **29**, 195-207.

Fisher, J.D., and Fisher, W.A. (1992). Changing AIDS risk behavior. *Psychological Bulletin*, **11**, 455-474.

Fisher, W.A., and Fisher, J.D. (1993). A general social psychological model for changing AIDS risk behavior. In J. Pryor and G.Reeder (Eds.), *The social psychology of HIV infection* (pp. 127-153). Hillsdale, NJ. Erlbaum.

Greenberg, B.S. (1988). Some uncommon television images and the drench hypothesis. In S. Oskamp (Ed.), *Television as a social issue* (pp. 88-102). Newbury Park,CA: Sage.

Hansen, C.H. (1995). Predicting cognitive and behavioral effects of gangsta rap. *Basic and Applied Social Psychology,* **16**, 43-52.

Hansen, C.H., and Hansen, R.D. (1988). Priming stereotypic appraisal of social interactions: How rock music videos can change what's seen when boy meets girl. *Sex Roles*, **19**, 287-316.

Holder-Nevins, D., and Bain, B. (2001). Popular music and sexual behavior among female adolescents in Jamaica-A case: control study. Special issue: HIV/AIDS and children in the English-speaking Caribbean. *Journal of HIV/AIDS Prevention and Education for Adolescents and Children*, **4**, 149-160.

Holmes, D. and Jackson, E. (2005). Wait (The Whisper Song) [Recorded by Ying Yang Twins]. On USA (united states of atlanta) [CD]. New York: TVT Records.

Huesmann, L.R. (1997). Observational learning of violent behavior. In A. Raine, P.A. Brennen, D.P. Farrington, and S.A. Mednick (Eds.), *Biosocial bases of violence* (pp. 69-88). New York: Plenum Press.

Huesmann, L.R. (1998). The role of social information processing and cognitive schema in the acquisition and maintenance of habitual aggressive behavior. In R. G. Geen and E. Donnerstein (Eds.), *Human aggression: Theories, research, and implications for social policy* (pp.73-109). New York: Academic Press.

Huston, A.C., Wartella, E., and Donnerstein, E. (1998). *Measuring the effects of sexual content in the media.* Menlo Park, CA: Kaiser Family Foundation.

Johnson, A. and Muchita, K. (1995). Shoot Ones, Pt. II [Recorded by Mobb Deep]. *On The infamous* [CD]. New York: RCA Records.

Johnson, J.D., Adams, M.S., Ashburn, L., and Reed, W. (1995). Differential gender effects of exposure to rap music on African American adolescents' acceptance of teen dating violence. *Sex Roles,* **33**, 597-605.

Jones, bin Olu Dara. (2000). Oochie Wally [Recorded by Bravehearts, Nas, and QB Finest]. *On QB finest* [CD]. New York: Columbia Records.

Jones, bin Olu Dara. (2001). Got Ur Self A Gun [Record by Nas]. *On Stillmatic* [CD]. New York: Sony Music.

Kim, J., and Rubin, A.M. (1997). The variable influence of audience activity on media effects. *Communication Research*, **24**, 107-135.

Kweli, Talib. (2002). Get By. *On Quality* [CD]. New York: Rawkus Records.

Lemieux, A. F. (2000). *Social influence on the social and aesthetic aspects of music preferences.* Unpublished Master's Thesis: University of Connecticut. Storrs, CT.

Metzler, C.W., Noell, J., and Biglan, A. (1992). The validation of a construct of high-risk sexual behavior in hetero-sexual adolescents. *Journal of Adolescent Research*, **7**, 233-249.

North, A.C., and Hargreaves, D.J. (2008). 'Problem music' and subcultures. In D.J. Hargreaves and A.C. North (Eds.), *The social and applied psychology of music* (pp.143-236). New York: Oxford University Press.

North, A.C., Hargreaves, D.J., and Tarrant, M. (2002). Social psychology and music education. In R. Colwell. and C. Richardson (Eds.*), The New Handbook of Research on Music Teaching and Learning* (pp. 604-625). New York: Oxford University Press.

Reyes, R.M., Thompson, W.C., and Bower, G.H. (1980). Judgmental biases resulting from differing availabilities of arguments. *Journal of Personality and Social Psychology*, **39**, 2-12.

Rubin, A.M. (1994). Media uses and effects: A uses-and-gratifications perspective. In J. Bryant and D. Zillman (Eds.) *Media effects: Advances in theory and research* (pp.417-436). Hillsdale, NJ: Erlbaum.

Srull, T., and Wyer, R. (1980). Category accessibility and social perception: Some implications for the study of person memory and interpersonal judgments. *Journal of Personality and Social Psychology*, **38**, 141-154.

Taylor, L.D. (2005) Effects of visual and verbal sexual television content and perceived realism on attitudes and beliefs. *The Journal of Sex Research*, **42**, 130-137.

Ward, L.M. (2003). Understanding the role of entertainment media in the sexual socialization of American youth: A review of empirical research. *Developmental Review*, **23**, 347-388.

Ward, L., and Rivadeneyra, R. (1999). Contributions of entertainment television to adolescents' sexual attitudes and expectations: The role of viewing amount versus viewer involvement. *The Journal of Sex Research*, **36**, 237-249.

Wester, S.R., Crown, C.L., Quatman, G.L., and Heesacker, M. (1997). The influence of sexually violent rap music on attitudes of men with little prior exposure. *Psychology of Women Quarterly,* **21**, 497-508.

In: Music: Composition, Interpretation and Effects
Editor: Tamara E. Ivanova, pp. 31-56
ISBN: 978-1-60876-170-8

Chapter 2

MUSICAL EMOTION ASSESSMENT, BRAIN CORRELATES, AND GENDER DIFFERENCES

José Luis Díaz, Enrique O. Flores-Gutiérrez, Yolanda del Río-Portilla and María Corsi Cabrera
Facultad de Psicología, Posgrado,
Universidad Nacional Autónoma de México, Mexico

Abstract

In the present contribution we review our understanding of musical emotion focusing in the hypotheses, approaches, and results obtained in our interdisciplinary group since 2001 when an initial study proved that different musical excerpts evoke a significant inter-observer agreement in the selection of previously organized emotion terms within a relatively homogeneous population of human subjects. We have used both a correlation measure among the different EEG electrode sites and metabolic images of brain function in non-musician human volunteers during the hearing of three instrumental music masterpieces selected for their capacity to generate pleasant or unpleasant music emotions. Using both fMRI and EEG coherence brain images during the listening of the three masterpieces, we found that music-induced emotions are sustained by an activation of specific brain sites and changes in the pattern of coherent activity between cortical regions. Both brain imaging approaches identified divergent brain networks for positive and negative music affective valence. A left cortical system involved with pleasant music feelings included the posterior temporal-parietal, occipital, and middle prefrontal regions. A coherent activity in upper alpha frequencies took place between most left hemisphere electrodes with a focus in the temporo-parietal and occipital regions. In other studies we found that the coherent alpha activity of the left hemisphere involved in pleasant music emotion is much larger in women linking all frontal regions with the left temporal and parietal-temporal association areas of the brain. Even though women and men reported similar levels of all emotions, the self-reported feeling of being happy was only significant in women and the number of significant adjectives of the *happy* component was more abundant in them. The predominance of left hemisphere activation with pleasant musical feelings and right activation with unpleasant ones is consistent with right hemisphere predominance for dealing with novel cognitive situations and the left for predictable representations. The emotional effects of music may have to do with symbolic and grammatical properties that engage representational semantic systems of the mind/brain. The activations of extensive networks gain emotional meaning in higher neocortical substrata intersecting with the language systems of the brain. The affective powers of music require higher symbolic mind/brain mechanisms developed and functioning in unison with cultural cosmologies. Music expresses and carries emotive and symbolic meanings capable of extending and transcending natural stimuli by impinging into deeply meaningful human emotions engaging semantic neocortical neuron networks.

The Musical Emotion Problem

Music is a humanly-produced sound construction which by means of accurately-adjusted instruments and improved motor expression skills becomes a spatiotemporally intricate auditory stimulus resulting in aesthetically meaningful conscious emotional states. Indeed, since time immemorial, music has been created, desired, and enjoyed mainly because of its powerful effects on emotions, feelings, and mood states (Balkwill and Thompson 1999; Juslin and Sloboda 2001; Baumgartner *et al.,* 2006). But even though music is traditionally considered a source of intense, diverse, and specific affective states, and despite the century-old debate about the emotional meaning of music, the scientific analysis of music emotion is a relatively recent, though vigorous undertaking (for an general review see Juslin and Sloboda, 2001).

In his classic "*Psychology of Music*" of 1939, the Swedish-American psychologist Carl Seashore (1967, p 9) declared that "Music is essentially a play upon feeling with feeling. It is appreciated only insofar as it arouses feeling and can be expressed only by active feeling". In accord with the behaviourist climate prevailing during the time of his writing, Seashore concentrated his analysis of music feeling in the expression of music considering feasible to develop a scientific analysis of all the behavioural elements of such expression. The counterpart of the emotions experienced by the listener is mentioned briefly in his book and only in terms of the individual differences in sensitivity that are not only innate but "intensified by pursuit". Such music experience may result in limited behavioural consequences since, beyond evoking impulses to sing, dance or swing to its rhythm while it lasts, attentive music listening not necessarily involves overt behaviour. In contrast to the typical and trans-cultural facial expressions of primary emotions induced by relevant social and ecological stimuli (Ekman, 1992), extremely stirring musical emotions usually implicate a rather limited facial gesticulation. Nevertheless, a robust field of music perception has been developing in the last decades but while a great deal has been learned concerning the sound and motor expression, and about many sensory and psychophysical aspects of music perception, not much is known about the affective experience of hearing music.

Several philosophers since Arthur Schopenhauer (1788-1860) have repeatedly suggested that emotions may constitute the semantic meaning of music, and that music somehow denotes or even embodies human emotion. In a bold and often quoted paragraph Schopenhauer wrote:

> "Music stands quite alone. It is cut off from all the other arts.... It does not express a particular and definite joy, sorrow, anguish, horror, delight, or mood of peace, but joy, sorrow, anguish, horror, delight, peace of mind themselves, in the abstract, in their essential nature, without accessories, and therefore without their customary motives. Yet it enables us to grasp and share them fully in this quintessence" (Taken from Bachelder, 1975, p 38).

In 1942 the symbolist philosopher Susanne Langer (1953) made a scholar and compelling allegation in this same sense and concluded that music does not only express or cause emotion, but it is *about* emotion, it is "the *formulation* and *representation* of emotions, moods, mental tensions and resolutions – a 'logical picture' of sentient, responsive life, a source of insight, not a plea for sympathy" (p. 188). In 1983, the composer and music theorist Fred Lerdahl with the language philosopher Ray Jackendoff, a former student of Noam Chomsky, carefully worked out the grammatical structure of music (the M-grammar) that following a set of unconscious rules allows the listener to consciously understand the piece. Diana Raffmann (1993) went a step beyond by affirming that we have musical experiences or feelings because the organized sounds generate these underlying M-grammar representations (p. 27) and that the meaning of a musical work consists in the feelings that result from the listener's unconscious recovery of structures constitutive of the listened piece (p. 53). But, since musical feelings are not easily put into words and conveyed by language, according to Raffmann they should be considered distinctive non-propositional representations of conscious awareness, not a form but an *analogue* of language semantics. Interestingly enough, Raffmann developed her music "quasi-semantics" disagreeing with the contention that music meaning is *about* emotions. She compellingly argued that primary emotions, such as happiness, sadness, anger, fear, contempt, and surprise cannot constitute the semantic sense of music since, in contrast with the content of these emotions when they arise after specific natural stimuli, the feelings during music hearing do not contain specific denotations and do not induce adaptive behavioural responses. Johnson-Laird and Oatley (2008) maintain in a similar manner that even though music can stir and produce powerful primary emotions, they are not necessarily *about them* in a linguistic or propositional sense. These authors clearly admit that we ascribe emotional properties to the music we hear and that such properties may coincide with the intentions of the composer and the performer of a piece. Thus, the communication of musical feelings would occur because there is a correspondence of musical representations in the minds of composer, performer, and listener (Raffmann, 1993, p. 55). Since such communicative properties have been reasonably well proven (see, for example, Juslin and Sloboda, 2001), it would be hard to discard that music constitutes a language of sorts even if the feelings were not always labelled with words and constitute qualitative sensations or connotations.

Not only the similarities between music and language have justified the notion of *music semantics* (Meyer 1956; Brown 2001) but Damasio (1989) and Mesulam (1998) consider the attribution of meaning to a stimulus of any modality as properly semantic. Indeed, the precisely-organized auditory stimuli considered musical require different brain modules and systems to be involved in distinct cognitive tasks including the meaning extraction of a nonverbal message (Mesulam 1998; Michel *et al.,* 2001; Newman 1997). In a chapter ingeniously entitled "The cognitive tango" the music professor Lawrence Zbikowski (2006) has shown that categorization, cross-domain mapping, and the use of conceptual models operate at basic levels in music understanding. Probably these and other major cognitive capacities are involved in the generation of musical emotions. Nevertheless the main difference between music and language is the lack of propositional content of instrumental music, and this fact makes not only a full comparison impossible but represents a challenge for music semantics. But if propositional content is a core feature of language, the wordless expression and creation of emotion is a central semantic element of instrumental music. And the fact is that we do not know how pure instrumental music realizes in the listener the amazing emotive effects that it does in the absence of representational content. As neuroscientists we are inclined to think that the answer to this fascinating mystery must have something to do with a correspondence between music spatiotemporal structures and brain/mind spatiotemporal processing, a correspondence having to do with the fact that both can be conceived and modelled as homologous and formally isomorphic psychophysical patterned processes (Díaz, 1997; 2007).

The idea that music merely triggers in the listener emotions felt in the past is contradicted by the fact that music is capable to elicit novel and peculiar emotional states. For example, the sadness produced by some pieces of music is not associated with the loss of a value or the frustration of an objective as it occurs in daily life, and it seems therefore remarkable that this and other musical feelings can be similar in kind and even greater in magnitude to naturally elicited emotions. The intriguing fact that people find aesthetic some negative musical emotions as if they were positive experiences is considered by Levinson (1997) a case of "mirror emotions" because music does not evoke the full-fledged emotion that follows particular ecological and social stimuli that are associated with beliefs, desires or attitudes. Since music emotions may occur without cognitive appraisal or content Levinson would probably agree with Raffmann (1993) in regarding musical feelings emotional *analogues* and not *proper* emotions. Davies (1997) considers that the enjoyment of negative emotions in art in general has to do with the possibility that what we find gratifying is a form of comprehension, a music epistemology, one may say, of "no pain no gain". In this sense it could be inferred that experiencing negative emotions outside the usual unsafe environment is a form of pretense not unlike that functioning during play or game behavior (Walton, 1990). All art is artifice and music is particularly endowed with mimetic powers to induce emotions. Thus, the emotions and moods of particular situations are metamorphosed into those movements of music that correspond most faithfully to such emotions. Why and how this happens is, at least in part, another theme for cognitive science and neuroaesthetics.

Aside from the difficulties posed by trying to understand the ontological and aesthetical status of musical emotion, the methodological problem of scientifically analyzing musical emotion is not only that the feeling state is essentially subjective but also that it is difficult to analyze the reasons why intrinsic music structure is responsible for the felt emotion. Some general principles have been discovered and employed for centuries by musicians and music scholars. For example, it is known since classical times in occidental music that mayor keys and rapid tempos cause happiness, that minor keys and slow tempos produce sadness and that dissonance produces anxiety and fear. Specific associations of particular emotions with particular aspects of musical composition have been outlined since the XVII Century. David Freedberg (2006) reproduces Marc-Antoine Charpentier's list of 17 key modes coupled with moods and feelings, such as C Major with "gay and warlike", E-flat Major with "cruel and severe", G Major with "quietly joyful" and so on. But, as Freedberg reminds us, the emotional impacts of music are infinitely more complex due to modulations of keys, to rhythm, harmony and melody, not to mention the listener's cultural, individual, and setting variations.

Another possible avenue to deal with these important philosophical and aesthetic problems posed by musical emotions is the empirical study of the human brain while a subject is processing musical stimuli and experiencing conscious emotive states that may communicate to the cognitive neuroscientist. The advent of noninvasive techniques to visualize brain dynamics has allowed for many complex cognitive states to be analyzed in terms of their neurological substrate and these results provide an important ingredient for the discussion as we shall briefly review now.

The Brain and Musical Emotion

In dealing with music emotion it must be taken into account that music must touch on something of the human mind/brain that brings people into a particular emotional state. How and why this happens is a central question concerning neuroaesthetics, an emerging interdisciplinary field situated between neuroscience and aesthetics (Zeki, 1999; 2003). In order to address such questions it is necessary to ponder and pose the distal question of how the evolutionary process shaped the human mind/brain to appreciate music (Darwin, 1872; Huron 2001; Fukui, 2001), and the proximal question on how the brain functions correlate with the emotions evoked by music (Panksepp and Bernatzky, 2002). Thus, one form of approaching the problem of musical emotion more objectively concerns cognitive neuroscience. But even though brain responses to experimentally-produced musical features, such as pitch and melodic stimuli (Zatorre *et al.*, 1994; Patterson *et al.*, 2002; Gagnon and Peretz 2000), tonal information (Zatorre 2001; Janata *et al.*, 2002), musical timbre (Halpern *et al.*, 2004), or rhythmic and temporal structure (Sakai *et al.*, 1999; Samson *et al.*, 2001) have been extensively studied, musical emotions display important obstacles to be analyzed and resolved. For example, at the present time there is not a standardized scientific procedure that discloses emotional processes and events evoked by music with reliable confidence. Indeed, the progress in the understanding of musical emotion crucially depends in the development of reasonable secure methods to record and analyze such peculiar and universally-sought affective process. As it will be described later, this particular problem has been addressed in our research group and the principles and results of the approach will be summarized in the next two sections.

Another difficulty concerns the selection of the proper stimuli to evoke particular musical emotions. Music consists of a temporal flow of organized auditory events and its understanding and concurrent emotions can only be achieved over long time windows, in the order of seconds to minutes. Since the intricate arrangement of musical features is what evokes an overall affective response (Hevner, 1936; Altenmüller *et al.*, 2002) there has been an increasing interest to use real music as experimental stimulus (Krumhansl, 2003). Indeed, the use of actual music seems not only pertinent but required to study musical emotion. Specifically, music masterpieces are characterized by a skilful handling of various sound resources and musical elements to obtain the expression of an idea that usually evokes defined affective states, and consequently are rightly suited to analyze the neural correlates of the emotional response to music (Ramos and Corsi-Cabrera, 1989; Ogata 1995; Blood and Zatorre, 2001; Altenmüller *et al.*, 2002).

At the present time there are mainly two types of noninvasive techniques to visualize neural function and generate brain images in conscious humans submitted to cognitive tasks: electrical and metabolic measurements. The electroencephalogram (EEG) is not only a suitable method to follow the moment to moment changes induced during prolonged periods of music listening, but also provides a temporal average of brain activity induced by the continuous flow of tones when an epoch of several seconds is analyzed (Bhattacharya *et al.*, 2001; Patel and Balaban 2000). For example, ongoing electrical activity recorded within the hippocampus changes with dissonant and unsettling musical intervals (Wieser and Mazzola 1986) and increased theta and decreased alpha activity has been found using pleasant musical pieces (Ramos and Corsi-Cabrera, 1989).

The neural correlates of musical emotion have recently started to be considered and empirically approached. Panksepp and Bernatzky (2002) reviewed the ways in which music may promote emotional and behavioral changes, including its effects on memory, mood, brain activity, and the autonomic responses correlated with the experience of 'chills.' Blood and Zatorre (2001) showed that intense stirring and thrilling musical emotions require limbic and paralimbic networks involving the integration of sensory and cognitive information. An integration of limbic structures with perceptual and cognitive regions was reported in non-musicians experiencing pleasant feelings elicited by the passive listening of unfamiliar instrumental music (Brown *et al.,* 2004; Koelsch *et al.,* 2006).

A dissociation of the left and right hemisphere participation during pleasant and unpleasant musical emotions can be assumed because positive and negative affects involve predominant activation of the left and right hemisphere respectively (Davidson 1992; Blood *et al.,* 1999; Herrington *et al.* 2005; Koelsch *et al.,* 2006). Moreover, using music stimuli, Gagnon and Peretz (2000) found left hemisphere advantage for pleasantness ratings of tonal and atonal melodies, and activations of left fronto-temporal regions with musically elicited positive feelings and of right anterior brain regions with negative feelings have been reported (Altenmüller *et al.,* 2002; Schmidt and Trainor 2001). Music emotion possibly emerges from the initial activation of brain regions directly involved in cortical music perception, and the subsequent activation of independent emotional systems of the sensory modality, aside from coherent cortical activations involved in the extraction of musical meaning.

In our experiments we have used both EEG and metabolic images of brain function in human volunteers during the hearing of music masterpieces selected for their capacity to generate pleasant or unpleasant music emotions. Specifically we have used an EEG technique to measure the correlation at time 0 between different brain sites with the consecutive generation of brain EEG synchronization images. It seems important to summarize the basic notions of this approach and its relevance to the study of musical emotions.

Neuron excitability depends upon the level of membrane polarization. The temporality between the information input and the state of the neuronal assembly of a particular brain locus or module is a crucial factor in the information processing and interchange. The diverse neuron systems and modules of the brain have a capacity to modify the general state of activation of each other by means of inter-modular excitatory or inhibitory influences that modify the level of excitability of its neurons within very short time periods. Moreover, the excitability of a particular neuron system can be modified bottom-up and top-down by global influences. The functional simultaneity of global and/or local inter-modular influences sets the function of specific neuronal systems in synchronized states to integrate the inputs and the outputs of each system. As a result of the synchronic modification of membrane potentials in neuron systems, electrical field oscillations are generated and they are recorded in the surface of the skull by means of the EEG. The temporal coupling or synchronization of such oscillations occurring among distant neuron systems or brain modules most likely reveals simultaneous inputs and/or global influences affecting two or more neuronal assemblies simultaneously. This simultaneity is manifested by the synchronization of their oscillations as revealed by the correlation at time 0. Therefore, in our experiments, we use a correlation measure among the different EEG electrode sites in the supposition that an increase or decrease in the temporal coupling during a particular musical emotion probably reflects global influences of neuron systems involved in such emotion. In this case it seems justified to pose that the coupling or uncoupling of the different neuron systems may be a physiological factor that contributes to generate the hedonic valence of the musical emotion.

On the other hand, metabolic techniques allow the imaging of changes in brain blood irrigation, molecular turnover and other physiological variables required in varying rates according to the activity of the different regions and centers of the brain. Even though these are indirect techniques of neuronal activity they provide a powerful spatial resolution, in contrast to the EEG correlation techniques that allow a powerful temporal but a weak spatial resolution. In some of the experiments that we will summarize later, we employed both techniques to analyze the processing of musical emotions.

The experiments involving functional magnetic resonance imaging (fMRI) were performed on a 1.5 Tesla Magnetic Resonance equipment. Functional images were acquired each 3 seconds in 7 contiguous 8 mm thick axial slices. Structural images were obtained for each subject and all MR data were transferred to an offline workstation, saved, and loaded into a SPM2 program for group statistics, after image registration and alignment were secured. All functional images were normalized to the standard provided by SPM. Functional signals were obtained to a hemodynamic response function and the statistical analysis of fMRI data were performed using an autoregressive model where the relevant *regressors* were each of the musical pieces in contrast against all others to detect the differences.

In the following sections of the present paper we will briefly review and analyze our present understanding of music emotion in non musicians focusing in the hypotheses, experimental approaches, and results that our interdisciplinary research team has been producing since 2001. We have been particularly interested in developing methods to evaluate music emotion, to analyze the brain mechanisms involved in these particular affective states, and to analyze differences between men and women. In the final discussion we shall also present our ideas and results in the frame of the lively cognitive, philosophical, and aesthetical discussions concerning musical emotions.

The Evaluation of Musical Emotions

Since the pioneering contributions of Kate Hevner (1936) to music psychology where she systematically measured emotional attributions to musical pieces by the use of a circle of adjectives, several studies have advanced the methods and techniques to assess the emotions produced by music. There are many difficulties still unresolved concerning the criteria to choose the musical pieces, the model and terms of emotion employed, the design of the experiment, the proper controls, and the relevant statistical tools to analyze the results (for an overview of this problem, see Juslin, 2005). A particular difficulty in this type of research is to properly distinguish whether music elicits emotional responses or expresses emotions that listeners recognize in the music. These two possibilities are, in fact, supported by two schools upholding an "emotivist" or a "cognitivist" position, perhaps an avoidable dichotomy.

In order to test and advance an experimental technique to evaluate and study human emotions evoked by music we designed a music evaluation experiment based on an inter-subjective agreement paradigm (Flores-Gutiérrez, 2001; Flores-Gutierrez and Díaz, 2009). The study intended to prove if different musical excerpts evoke a significant inter-observer agreement in the selection of previously organized emotion terms within a relatively homogeneous population of human subjects. Since music constitutes a form of acoustic expression that has been selected and developed during millennia of human cultural evolution for the manifestation and communication of emotional states, it was expected that there will be a significant agreement in the attribution of terms of emotion to musical segments among human evaluators belonging to a relatively homogeneous population. The attribution system, which included a developed scheme of emotions, the attributions data obtained in groups of subjects submitted to carefully-selected musical stimuli, and the use of an inter-observer agreement statistical processing, allowed objective responses derived from introspection to be obtained.

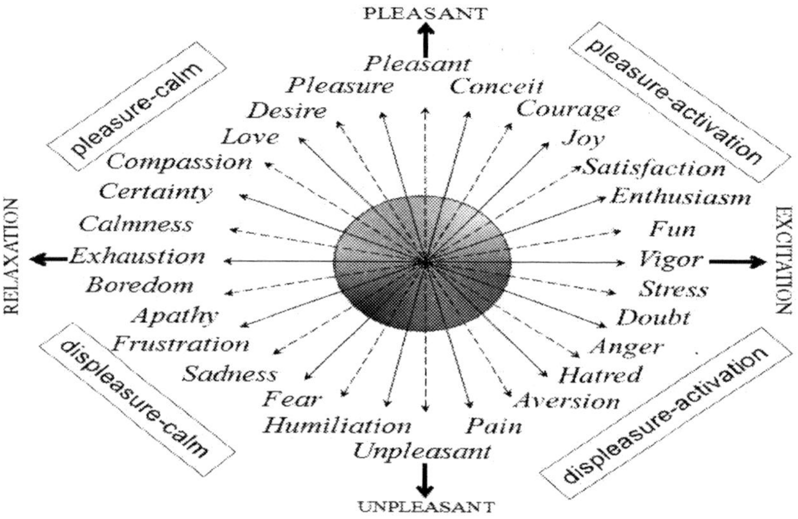

Figure 1. Circular model of the affective system. Fourteen polar axes of antonymic emotional concepts (of opposite affective sign) are placed in an orthogonal plane defined by two dimensions: a level of activation in the abscissa (excitation to the right, relaxation to the left) and a level of hedonic value in the ordinate (pleasant above and unpleasant down).

The volunteer subjects were 108 college-level non-musicians of both sexes with a mean age of 22 years belonging to schools and universities located in Central Mexico. The audition and attribution sessions lasted for 90 min and were conducted in a specially adapted classroom located in each institution. Four criteria were established for the selection of the musical excerpts: instrumental music, homogeneous melody and musical theme, clear and distinct affective tone, and samples of different cultures. The ten selected pieces, in the recorded order, were: (1) Mozart's piano concerto no. 17, K 453, third movement; (2) A sound of the magnetic spectra of an aurora borealis, a natural event; (3) Mussorgsky's *Gnome*, from *Pictures at an Exhibition* orchestrated by Ravel; (4) Andean folk music; (5) Tchaikovsky's *Fifth Symphony*, second movement; (6) "*Through the Never*", heavy metal music by Metallica; (7) Japanese *Usagi* folk music played with *koto* and *shyakuhachi*; (8) Mahler's *Fifth Symphony*, second movement; (9) *Taqsim Sigah,* Arab folk music played with *kamandja,* and (10) Bach's *Inventions* in three parts for piano, BMW 797.

The selected fragments were played twice during the test. During the first audition, the complete piece was played in order for the subjects to become familiar with the composition and freely express their reaction in writing. During the second hearing the same piece was played in the separate-selected segments and the volunteers were asked to choose those emotion-referring terms that more accurately identified their music-evoked feelings (not their attributions to what the music intended to express) from an adjunct chart obtained and arranged from an original list of 380 Spanish words designing particular emotions.

The emotion adjectives had been previously arranged in 28 sets of semantically related terms located in 14 bipolar axes of opposing affective polarity in a circumflex model of the affective system (Díaz and Flores, 1997; Díaz 2007). This model was based on a constructive analysis of a Spanish emotional vocabulary of six stages: a selection of a Spanish vocabulary of 380 terms denoting particular emotions, the grouping of these terms in 28 clusters or semantic fields, the selection of a representative term for each cluster, the ordering of the words in a sequence according to the intensity of the designed emotion, the identification of antonymous pairs of sets and, thereby, the establishment of 14 bipolar axes, and finally the elaboration of a circumflex model of the human affective system where the axes were located in reference to the orthogonal dimensions of hedonic value and activation. A translation of this scheme into English is presented in figure 1. The resulting model consists on a topographic circle of the affective system allowing for the allocation of individually-labelled emotions. The orthogonal intersection of these two variables produced four quadrants of emotion: upper left pleasure-activation, upper right pleasure-calm, lower left displeasure-activation, and lower right displeasure-calm.

With this scheme at hand the subjects recorded their attributions while listening to each music excerpt and their selections were captured and transformed into ranks. The non-parametric Friedman test of rank bi-factorial variance for k related samples was selected for the statistical analysis of inter-rater agreement. All the data were gathered in the 28 categories or sets of emotion obtained in the previous taxonomy of emotion terms and the difference among the musical segments was tested. The difference was significant for 24 of the 28 emotional categories for $\alpha = 0.05$ and 33 degrees of freedom ($F_r \geq 43.88$). In order to establish in which segments were the main significant differences, the extension of the Friedman test for comparison of groups to a control was undertaken. Thus, after applying the appropriate formula, a critical value of the difference $| R_l - R_u |$ was established at ≥ 18.59. In this way it was possible to plot the significance level of all 28 emotion categories for each music segment and thereby to obtain the emotion profile of each selected music fragment. A histogram for each piece was constructed by ordering the 28 emotion terms in the abscissa starting with vigor, so that the four quadrants appear in a linear order. The ordinate corresponded to the degree of attribution of each group of emotions.

The differences obtained for the musical pieces were established both for the significant response of individual emotion, groups of emotions, and the global profile of the response. In all the pieces used, one or more of the 28 terms showed statistical significance. Sometimes as many as 7 terms appear predominant (Mahler, Mozart). In contrast other segments produce only one or two responses (Aurora borealis, Arab music). In most musical segments there were null responses implying that there was an agreement not only concerning the emotions that were present, but also those that did not occur at all.

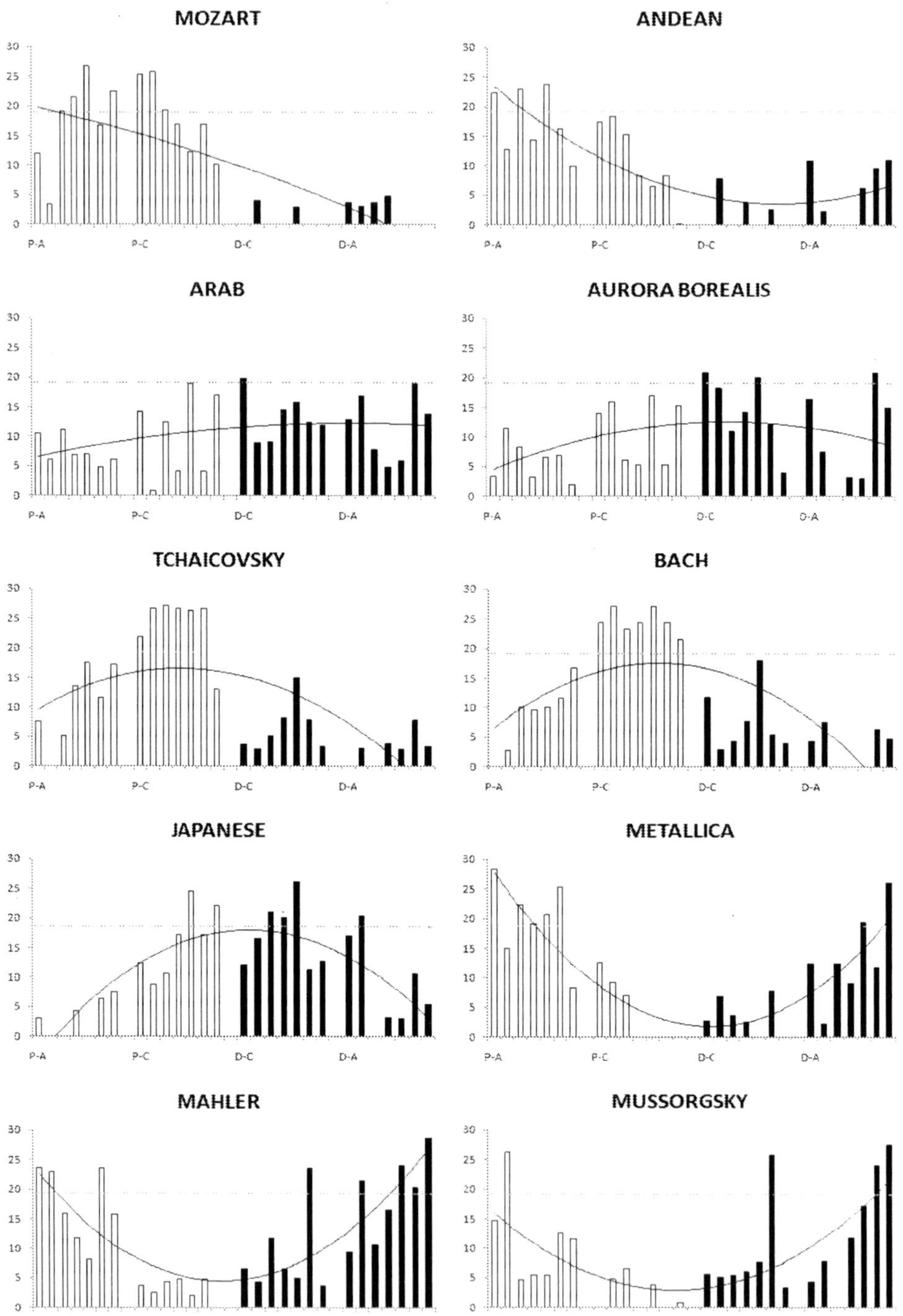

Figure 2. Subjective evaluation of ten musical stimuli. Representative graphs for each one of the ten musical pieces. Subjective evaluation data were gathered in the 28 categories or sets of emotion defined in fig. 1. The histogram for an excerpt of each piece is constructed by ordering the 28 emotion terms in the abscissa starting with vigor. The ordinate corresponded to the degree of attribution of each group of emotions. A significance threshold was established at ≥ 18.59 (Freedman test). Evaluation of the Emotions Induced by Three Music Masterpieces.

Concerning the global response there were several profiles recognizable among different pieces (Figure 2). In this way, the histogram is slanted to the left when positive and vigorous emotions are reported (Mozart, Andean music). The predominance of emotions in the center-left section corresponds to positive and quiet emotions (Tchaikowski, Bach), the center-right sector corresponds to negative and quiet emotions (Japanese music) and the fourth sector of negative and agitated emotions (Mahler). Sometimes a "U" shaped profile was obtained when vigorous emotions predominated (Mahler, Metallica). A bell-shaped response was obtained when calm emotions were reported, both pleasant and unpleasant (Japanese music). There is also music that globally stimulates one of the four axes defined in the affective circle, such as pleasant (Mozart), unpleasant (Mussorgsky), exciting (Metallica) or relaxing emotions (Japanese music). The only piece that produced scattered responses in the four sectors of emotions was the aurora borealis. Very similar profiles were obtained with very different pieces, such as the nearly identical responses to Mozart and Andean music and it would be necessary to analyze the individual emotion terms to distinguish them. Several common characteristics can be detected in these two pieces, such as fast speed in tempo allegro, binary rhythm, counterpoint figures, and ascending melody, well known features in music composition to evoke joy. In contrast, other segments evoked unpleasant responses (Mussorgsky), where *fear*, *tension*, *doubt* or *pain* was reported. As it was mentioned earlier, the listener probably concedes a high discernment or noetic value to a piece that evokes emotions that normally avoids in the context of a controlled and safe artistic experience.

If the terms of emotion selected by the volunteer subjects actually corresponded to relatively discrete emotional states, most musical segments appear to evoke an emotional response that is similar among the subjects due to their composition and interpretation characteristics. In other words: musical segments selected for their expressive characteristics seem to generate an emotional response which is similar among listeners with comparable cultural and educational backgrounds. The cultural factor seems to have influenced the results, since the responses were more defined and robust with familiar types of music so that Japanese and Arab music evoked the least distinctive and vigorous responses. It would be of interest to perform the same experiment in Arab and Japanese subjects to compare the response.

Even though the previous technique proved to be potentially useful to generate and analyze specific emotional status in experimentally-controlled situations of musical audition, the method used in successive cognitive neuroscience and brain image experiments had to be initially restricted to pleasant, unpleasant tranquil and energetic feelings because of the time limitations of the techniques employed. In order to correlate the feelings with brain images, visual analogue scales were used. In this section of the article we shall summarize the procedure to evaluate such musical feelings and that was used in all the succeeding metabolic and EEG brain image studies. As it will be shown, a validated subjective evaluation test using multiple feeling terms, and a Principal Component Analysis of such assessment, clustered two opposite emotional poles (pleasant and unpleasant feelings), and an activation profile. Once this objective was properly accomplished, it was justified to compare the metabolic and electrical brain correlates to each one of these three significant affective factors.

Two pieces of music were selected for inducing happiness, a soft piece for piano in tempo andante *(Invention for three parts, BWV 789* by J.S. Bach*)*, and an energetic and dramatic piece for symphony orchestra (the first part of the second movement *allegro-agitato*, of the *Symphony number 5* by G. Mahler). In accord with another study from our group performed in 108 students (Flores-Gutierrez, 2001), both pieces were equally strong for inducing happiness but differed in their capacity for inducing activation. For the induction of fearsome emotions we used a disturbing and scary piece of music written by the contemporary composer J. Prodromidès for the film *Danton* which was previously validated as fear-inducing by 335 subjects (Ramos *et al.,* 1996). Furthermore, aside from their opposite valence, the Danton piece by Prodromidès and the music by Mahler elicited equal levels of activation. In accord with the previously mentioned topographic model of the affective system (Díaz and Flores-Gutiérrez, 2001) we used musical pieces that produced emotions located in three of the four quadrants formed by the orthogonal intersection of the hedonic valence and activation variables of emotion: pleasure-activation (Mahler), pleasure-calm (Bach), and displeasure-activation (Prodromidès).

The tracks were edited as required by means of the Sound Forge 5.0 program and digitally recorded on a CD. Since the comprehension and affective responses to music require a psychological integration over long periods of time, acoustic stimuli were presented for 10 continuous consecutive lapses of 30 seconds of the same musical piece, alternating with 10 lapses of 30 seconds of random noise constituted by a flat and homogeneous spectrum from radio static. The order of presentation of three musical excerpts was counterbalanced among subjects. The subjects were comfortably seated with their eyes closed inside a sound-attenuating chamber under dim light and the tracks were played in two loudspeakers located 2 m from the subjects.

Subjective emotional feelings elicited by each musical piece were evaluated by visual-analogue scales of 17 adjectives corresponding to different emotions previously validated in two independent groups of 215 and 80 + 40 (120) subjects (Ramos *et al.,* 1996). Thus, each adjective scale was a 10 cm horizontal bar representing absence of the emotion at the left and highest grade at the right. Three yes/no questions explored weather the subjects *liked* the piece, *would like to hear it again*, and if they *knew it*. Only the subjects that did not know the musical pieces were included in the study.

Quantitative results from subjective scales were submitted to principal component analysis (PCA) (Jolliffe, 1986) and the resulting component scores of each eigenvector were compared with mixed ANOVAs and the music pieces as the within-subjects variable. *Post-hoc* comparisons with Tukey´s t-test allowed to determine differences associated with subjective feelings for each musical piece. A significance level of $p < 0.01$ was established after Bonferroni correction.

Five components or factors were identified by PCA (table I) which were very similar to those found in previous studies (Ramos *et al.,* 1996; Flores-Gutiérrez, 2001). *Emotional valence*, *activation*, and *attention* were identified in separate factors and the same three valence-related components were recognized, one for *happiness*, another for *fear* and a third for *sadness*-related emotions. The feeling of being happy was related with being pleased and strongly co-varied with the feelings of being calm and comfortable, whereas the emotion of fear was related with being uncomfortable and covaried with the feelings of being tense and restless. The hedonic value (pleasant- unpleasant) of the Bach and Mahler pieces used was also corroborated by the answers to the first two yes/no questions. Most of the subjects *liked* each one of the two pieces ($p < 0.01$) and *would like to hear it again* (p 0.05). Only happy and fear emotions were differentially elicited by the three music pieces. Although properly identified and separated from happiness and fear, sadness was not especially elicited by our stimuli. The same level of happy emotion was reported by two of the music pieces (Bach and

Mahler). In both cases all the scales involved in the *pleasant* component were rated as significantly higher and all scales of the *unpleasant* component as significantly lower. Conversely, *fear* was elicited only by Prodromidès' music with significantly higher scores in all adjectives of the *unpleasant* component and lower scores in all adjectives of the *pleasant* component. As expected, the self-reported feeling of *activation* was differentially elicited by the two music pieces eliciting happy emotions (Bach and Mahler) while it was similar for one of the happy pieces (Mahler) and by the fear-eliciting music (Prodromidès). All three musical stimuli elicited a similar intermediate level of *attention*.

Table I.

EMOTION ADJECTIVES		FACTORS Happiness	Fear	Attention	Sadness	Activation
ALEGRE	Cheerful, Glad	**.83**	-.17	-.02	-.32	0
ANIMADO	Lively, Spirited	**.61**	-.08	.16	-.14	**.57**
COMPLACIDO	Satisfied, Content	**.73**	-.31	.12	.19	.04
CONFORTABLE	Comfortable, Placid	**.76**	-.13	.31	.11	-.17
ENCANTADO	Delighted, Charmed	**.83**	-.14	.04	.01	.11
FELIZ	Happy,Joyful	**.86**	-.08	-.14	-.28	-.02
INSPIRADO	Inspired, Exalted	**.52**	-.30	.24	.50	.36
TRANQUILO	Calm, Quiet	**.63**	-.27	.08	.02	**-.56**
ASUSTADO	Frightened, Scared	-.43	**.55**	-.12	.20	.36
ENOJADO	Angry, Displeased	-.09	**.58**	-.28	.17	.16
FASTIDIADO	Annoyed, Bothered	-.35	**.65**	-.49	-.04	.04
INCÓMODO	Uncomfortable, Uneasy	-.41	**.61**	-.40	-.01	.11
INQUIETO	Restless, Distressed	-.27	**.82**	-.03	.15	.22
TENSO	Tense, Anxious	-.41	**.75**	.06	.09	.28
ATENTO	Attentive, Mindful	.12	-.02	**.91**	-.03	.03
INVOLUCRADO	Involved, Interested	-.04	-.07	**.81**	.17	.25
AFLIGIDO	Afflicted, Grieved	-.30	.43	-.10	**.66**	-.05
TRISTE	Sad, Sorrowful	-.02	.11	.13	**.91**	.00
EXCITADO	Excited, Roused	.04	.21	.20	.04	**.82**
Eigenvalues		5.12	4.66	2.28	1.94	1.89
% of Variance Explained		24.40%	22.19%	10.84%	9.23%	9.00%
p (Bach vs Prodromidès)		.0026	.0014	-----	-----	.0011
p (Mahler vs Prodromidès)		.0039	.0057	-----	-----	-----

Note: Varimax rotated values from subjective scales submitted to Principal Component Analysis. All eigenevalues higher than 1 are shown; factor loadings higher than 0.5 appear in boldface. Significant differences at p <.01 (Wilcoxon test) with music pieces as within-subject variables are shown..

Subjective evaluations of the three musical stimuli were similar to those previously obtained by Ramos *et al.*, (1996) and Flores-Gutiérrez (2001). The masterpiece selection was based on these studies where the distinction between emotional terms was based on meticulous tests. These results justified four types of experimental analyses: (1) pleasant musical emotion (Mahler and Bach vs. Prodromidès), (2) unpleasant musical emotion (Prodromidès vs. Bach and Mahler), (3) musical "activation" in terms of excitement (Mahler vs. Bach) or calmness (Bach vs. Mahler), and (4) general musical structure features not including specific emotions (Bach + Mahler + Prodromidès vs. noise).

Brain Correlates of Pleasant and Unpleasant Music Emotions

Using both fMRI and EEG coherence brain images, we recently reported that music-induced emotions are sustained not only by activation of specific brain sites, but also by changes in the pattern of coherent activity or simultaneously synchronized oscillations between cortical regions (Flores-Gutiérrez *et al.*, 2007). In this study brain correlates of some emotional reactions to music were analyzed in non-musicians during the listening of the three unfamiliar instrumental masterpieces without the request to perform any cognitive task, except paying attention to the music. In accord with the results described in the previous section, the piano excerpt by J.S. Bach and the passage from G. Mahler's 5th Symphony were chosen because consistently elicit pleasant emotions, and a segment by J. Prodromidès because it was reliably reported to engender unpleasant emotions. Brain correlates of emotional experience were singled out by comparing the pleasant and unpleasant emotional states reported by the subjects after listening to these music excerpts. Brain image correlates of pleasant music feelings were assessed by subtracting the periods of Prodromidès listening from the average of Bach and Mahler pieces and from each piece separately. Unpleasant feelings were evaluated by the opposite procedure. General and emotionally-unspecific music processing was evaluated by subtracting the noise periods from the average of the three music pieces.

The control of music characteristics such as structural features, expressive sequences, and cultural or idiosyncratic norms was accomplished by adding and comparing musical samples sharing an emotional effect but widely differing in musical features. Thus, the pleasant feelings elicited by the first two very different pieces were compared to each other and together to the unpleasant feelings elicited by the third dissonant piece. The strategy of using music pieces widely different in melody, rhythm, harmony, intensity, spectrum, or instruments used, but evoking similar emotional effects constitutes a rigorous control of these fundamental musical variables when their combined effects are detected in the brain. Moreover, since none of these pieces is commonly played in popular media and our subjects stated that they did not know them, such strategy provides a further control of the cultural and learned effects.

The pleasant or unpleasant responses to these carefully selected music excerpts were expected to correlate with distinctive images of localized brain metabolic activation and patterns of coherent interactions among brain regions. Even though the general metabolic and coherent EEG responses was explored without a-priori defining regions of interest, we expected to find predominant left hemisphere activation with pleasant music emotions and right hemisphere activation with negative musical emotions coupled with a distinct limbic and paralimbic involvement.

The coordinated application of standardized subjective evaluations, fMRI, and analysis of EEG coherent activity offered adequate means to reveal both brain sites (fMRI) and functional relations (EEG coherent activity) underlying some musical emotional experiences. Listening without any cognitive task except paying attention to the music engaged a widespread and diverse activation of brain sites and networks. Different brain systems were involved in the perceptual and affective processing of the selected pieces, and particularly in the processing of the positive and negative musical emotions.

Even though a correspondence between EEG and fMRI data could not be expected because they measure different physiological processes, we assigned special relevance to the areas where metabolic and electric correlation matching occurs. In effect, both fMRI and EEG coherence approaches identified divergent brain networks for positive and negative music affective valence. A left cortical system involved with pleasant music feelings included the posterior temporal-parietal, occipital, and prefrontal regions. A coherent activity in upper alpha frequencies took place between most left hemisphere electrodes with a nodal focus in the temporo-parietal and occipital regions indicating a functional coupling among some of the activated regions identified by fMRI.

The activation of the left primary auditory cortex by pleasant music was an unexpected finding of great consequence because it would imply that musical emotion is processed upon arrival of the stimulus at the earliest cortical location, traditionally considered to discriminate only for fundamental sensory features. Even though, as expected, all three music stimuli activated the superior temporal gyrus in both hemispheres, only pleasant musical emotions involved the left gyrus and only unpleasant ones the right gyrus. It seems feasible that the recognition of particular perceptual music features is prone to induce pleasant emotions. Pleasant affective musical processing may involve auditory areas in the dominant hemisphere providing some musical stimuli with a primary positive valence while unpleasant music involves the right auditory areas contributing to an initial attribution of negative emotion. Specific feelings elicited by music require the extraction of a meaning involving subsequent cortical locations. The activation and functional coupling among posterior temporal, inferior parietal and prefrontal regions seem to be needed to accomplish this task. Unpleasant emotions were associated with an increased coherent activity among both prefrontal and midline electrodes towards the right frontal and left temporal regions disrupting coherent activity with posterior regions in the left hemisphere. The activation areas involving right cortical regions found only for unpleasant emotions both with fMRI and EEG are important constituents of the paralimbic system.

The predominance of left hemisphere activation with pleasant musical feelings and right activation with unpleasant ones is consistent with findings that relate right frontal activation with negative affect and left frontal activation with positive affect (Davidson 1992). This result is consistent with right hemisphere predominance for dealing with novel cognitive situations and the left hemisphere for predictable representations and strategies, such as language (Podell *et al.*, 2001). Bach's and Mahler's are masterpieces that follow melody and harmony rules leading to a fulfillment of semantic expectations. In contrast, the piece by Prodromidès breaks classical musical "grammar" rules probably demanding the participation of limbic and right hemisphere regions in the search for emotional comprehension. In addition to the bilateral auditory sensory areas, brain regions involved in working memory, attention, and language processing were recruited in non-affective perceptual responses as determined by comparing all three musical pieces against noise.

The Type and Number of Emotions Induced by Music Masterpieces Differ between Men and Women

Convergent evidences from clinical, psychological, and neuroimaging studies show that emotional response is different in men and women. Women suffer more often from anxiety disorders and depressive illness than men. (Pigott, 1999); they are faster than men in recognizing non-verbal emotional cues such as prosodic information (Hall, 1978) and facial expression of emotions (Mandal and Palchoudhury, 1985). Moreover, women suffer stronger interference from incongruent language and prosodic information and integrate both signals faster than men (Schirmer and Kotz, 2003). Some studies have shown sex differences in reference to emotional and perceptual aspects of music. Not only brain responses to music are different in men and women (Schmidt and Trainor, 2001; Koelsch *et al.,* 2003) but also, compared to men, women prefer soft music to heavy metal and hard rock (Christensen and Peterson, 1988), and show greater positive attributions to music (Altenmüller *et al.,* 2002).

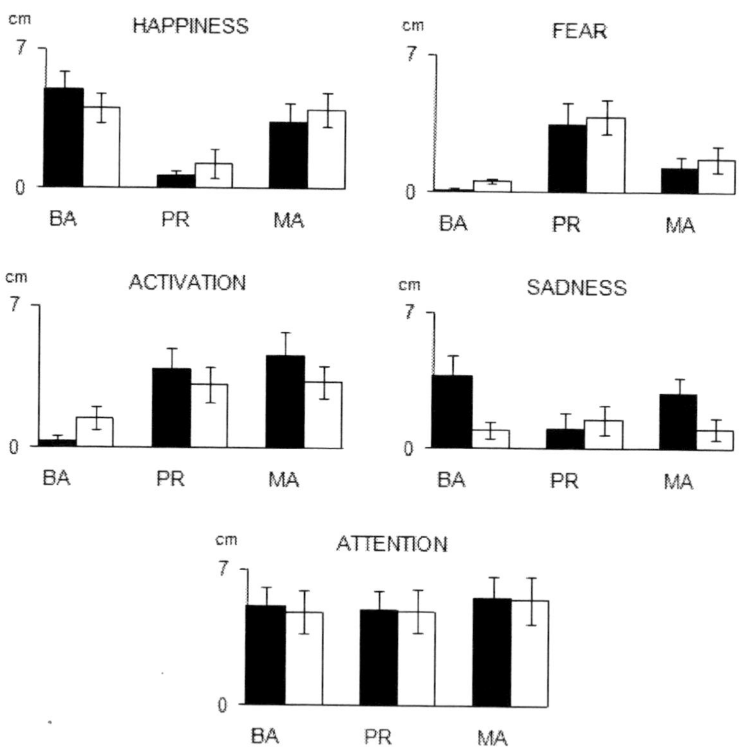

Figure 3. Factor loadings of emotional adjectives evaluating the subjective effects of music excerpts. Principal Component Analysis among musical pieces gathered a "happiness" component (eigenvalue: 5.12) in the left column and a "fear" component (eigenvalue: 4.65) in the second column. Bars represent rating means (±SE) of the adjectives elicited by the three musical excerpts in men (white bars) and women (black bars). The differences were not significant in the components of "sadness" and "attention".

These evidences suggest that the subjective evaluation of emotions evoked by music can be expected to differ between men and women. For this purpose, subjective ratings of 17 emotions elicited by music were compared between men and women using previously validated musical pieces and visual analogue scales of emotion terms. After giving written informed consent, 16 right-handed non-musician students (8 men and 8 women, mean age 25 years ± 3.05 SD) from the university community participated in the study as passive music listeners. In accord to the subjective emotions elicited by them, the three musical excerpts described above by Bach, Mahler and Prodromidès that were unknown to the participants were chosen as stimuli. Student-t tests comparing self-reported feelings for each subjective scale were done separately for men and women. A p level < 0.05 was required as significant. Yes/no questions were compared by chi square test

Table II. Emotional subjective response to music excerpts in women and men

		WOMEN	MEN
	BEING HAPPY		
alegre	**(cheerful, glad)**	0.024981	0.139250
animado	**(lively, spirited)**	0.890667	0.702324
complacido	**(satisfied, content)**	0.000007	0.094805
confortable	**(comfortable, placid)**	0.018914	0.016699
encantado	**(delighted, charmed)**	0.010044	0.602096
feliz	**(happy, joyful)**	0.027702	0.377596
inspirado	**(inspired, exalted)**	0.020872	0.237739
tranquilo	**(calm, quiet, pleased)**	0.001981	0.000239
	BEING SCARED		
asustado	**(scared, frightened, startle**	0.099372	0.012749
enojado	**(angry, displeased)**	0.174557	0.428586
fastidiado	**(annoyed, bothered)**	0.099544	0.070585
incómodo	**(uncomfortable, uneasy)**	0.074673	0.031710
inquieto	**(restless, distressed)**	0.043471	0.147209
tenso	**(tense, anxious)**	0.007186	0.015098
	BEING ACTIVATED		
atento	**(attentive, mindful)**	0.814354	0.548565
involucrado	**(involved, interested)**	0.640102	0.549483
afligido	**(afflicted, grieved)**	0.411066	0.934432
triste	**(sad, sorrowful)**	0.259093	0.904288
excitado	**(excited, aroused)**	0.014686	0.079996

▢ = $p < 0.05$

Note: Emotion adjectives were provided to subjects in Spanish, the English equivalents are shown in bold characters. Feeling-denoting adjectives in the first column were rated by 7 men and 7 women using visual analog scales. Numbers represent (p) values of the Student comparison *t*-test between the pieces by Bach-Mahler (pleasant) and Prodromidès (unpleasant) for women and men.

After listening to the three musical stimuli, women and men reported similar levels of all emotions, including happy and fear emotions, attention, and activation (figure 3). However, as shown in table II, despite that the intensity of emotional experience was moderate and similar in men and women, the self-reported feeling of being happy was only significant in women and the number of significant adjectives of the *happy* component was more abundant in them. Both, men and women were pleased and comfortable but only women felt they were happy, cheerful, delighted and inspired. In contrast, only men reported being significantly scared in addition to tense and uncomfortable, while women reported only being restless and tense in the *fear* component. The feeling of being activated was only significant for women who felt more aroused.

Thus, after listening to three music masterpieces widely differing in their capacity to evoke emotions, men and women reported the same level of being happy and scared and showed a general agreement for the valence-related feelings to these two emotions as identify by PCA. However, being happy and cheerful together with being pleased was reported only by women while men identified being pleased only by being comfortable and tranquil. Conversely, only men experienced fear together with being unpleased. These results are consistent with female preference of soft music (Christensen and Peterson, 1988) and greater positive attributions to music than males (Altenmüller *et al.*, 2002). The results are also consistent with women being better in recognizing facial expressions of happiness than men (Campanella *et al.*, 2004).

The more significant self-reported fear in men is also consistent with larger size (Goldstein *et al.*, 2001) and greater activation of the amygdala and temporo-limbic structures (Gur *et al.*, 2002) which are preferentially activated by fear (Schienle *et al.*, 2005). Finally, the self-reported happiness in women is consistent with larger size (Goldstein *et al.*, 2001) and greater activation (Andreason *et al.*, 1994; Gur *et al.*, 1995) of orbitofrontal and cingulated cortices which are known to exert an inhibitory action on the amygdala.

The higher sensitivity of women to happy emotions and the higher sensitivity of men to fear emotions elicited by music is consistent with an evolutionary perspective of music acquisition and development. In 1872, Darwin suggested that music might have arisen due to a sexual selection of mating calls creating a competition for more elaborate and pleasing melodies in males. In his view, music is a too complex and peculiar biological adaptation to have occurred except through direct selection for some survival or reproductive advantage. Since no survival benefits for music were articulated, Darwin emphasized that complex and creative acoustic displays are outcomes of sexual selection and function as courtship displays to attract sexual partners. In a different elaboration of the evolutionary hypothesis, Fukui (2001) suggested that music originated with the need to control the prevalently male sexual and aggressive behavior by easing strain, strengthening social bonds, and bringing cohesiveness to the group, that are predominantly female functions. If this is the case, women would be expected to be more sensitive to the pleasing aspects of music while men were more sensitive to fear.

Differential EEG Correlation Patterns in Men and Women Experiencing Pleasant and Unpleasant Musical Emotions

Since the functional relationships among brain sectors play an important role in cognition, emotion, and music processing, we investigated the electrical coherence or synchronization of brain areas in a group of young men and women during their listening of the three music masterpieces. In this study (Flores-Gutiérrez, *et al.*, 2009) we found that musical emotion entails particular interchanges among brain regions and the predominantly coherent alpha activity of the left hemisphere involved in pleasant music emotion is much larger in women. Thus, even though pleasant emotion involves a link between posterior association areas and frontal regions of the left hemisphere in both men and women, it seems clear that the coherent brain network is wider in women. The results also indicate that different regional couplings intervene for pleasant and unpleasant musical emotions. These findings agree with consistent brain differences between the sexes that correlate with their differential emotional responses, including their response to music.

In the design of the study we took advantage of two apparently diverse facts pertaining to both aesthetics and neuroscience. The first is that music masterpieces are skilful, polished, and deliberate expressions of intricate sound signals designed to communicate and induce defined emotions. The second is that electrical synchrony among two or more brain areas within a similar rank of electrical oscillations indicates an enhanced functional relationship involved in information processing.

In the experimental paradigm described above, the EEG activity was recorded using a standard scalp electrode arrangement in seven men and seven women of our sample during the music and noise listening periods. The signal was amplified and digitized so that the correlation spectra between pairs of derivations could be obtained every two seconds. The particular interchanges among brain regions were analyzed by means of their couplings in the alpha EEG frequency (8-12 Hertz or cycles per seconds), a brain rhythm known to occur during several sensory, cognitive and emotional processes. The significantly increased or decreased relationship among brain sites was obtained from an average of at least 154 seconds per subject providing a robust measure of their involvement in specific emotions.

The correlation analyses revealed different patterns of functional and temporal coupling between brain regions associated to subjective pleasant and unpleasant emotions in specific frequencies. As it was foreseen, different coherent networks were present in men and in women (Figure 4). Pleasant emotions elicited by the pieces of Bach and Mahler were sustained by coherent oscillations in the left hemisphere in both sexes, but by a particularly larger network in women. This extensive brain network of coherent activity in women linked all frontal regions with the left temporal and parietal-temporal association areas of the brain. Unpleasant emotions were supported by an increased coherent activity between midline and posterior regions in the right hemisphere in men and by bilateral networks engaging anterior regions in women.

The increased functional coupling of temporal and parietal regions of the left hemisphere in both men and women during pleasant musical emotions could be expected because the left hemisphere is known to be more involved in positive emotions. But the results showed specific and remarkable sex differences in lateralization patters. Thus, whereas in men the left hemisphere participates in pleasant emotions and the right in unpleasant emotions, the predominantly bilateral participation in women and the right one in men may be related to women's lower hemispheric specialization and greater propensity to be invaded and engaged by emotion. Finally, since the combined music effects showed increased upper alpha coherence in women and of low alpha activity in men it is possible that music processing involves more working memory in women and more selective attention mechanisms in men. The results indicate that men and women should not be analyzed together in neurobiological studies of emotion, particularly those involving musical stimuli.

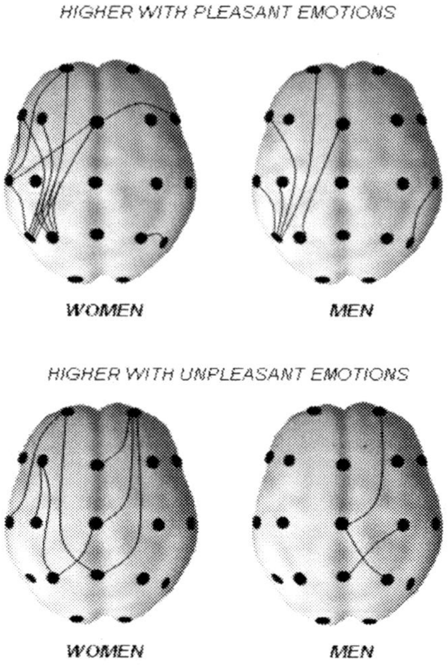

Figure 4. EEG coherent activity during pleasant and unpleasant musical emotions. Lines connecting electrodes indicate significantly alpha band coherent EEG ($p < 0.005$). Pleasant and unpleasant emotions for women (left side) and men (right side). See Table 1 for corresponding adjectives and significant level. Location of the electrodes is depicted in the surface of the average brain (Evans et al., 1993) as employed by Okamoto et al. (2004)

Discussion

Music probably derives its affective potency from the dynamic processes of brain systems that normally control "real" emotions and which are highly integrated with cognitive processes. Such systems do not only include subcortical and limbic structures, that have been shown to be involved in basic emotions and are affected by music (Blood and Zatorre, 2001; Panksepp and Bernatzky, 2002), but our results strongly highlight neocortical structures usually associated with cognitive processes as predominant systems involved in musical emotion. The involvement of these areas, particularly the left hemisphere of the brain during the experience of pleasant musical feelings, suggests that much of the emotional properties and effects of music have to do with a symbolic and grammatical property of music that, in similarity with language, engages representational semantic systems of the mind/brain.

Our results contradict to some extent the idea that the right hemisphere provides a predominant prosodic element in music expression and appreciation while the left hemisphere is involved in analytical aspects of music that are developed during formal music training. Our subjects were not trained in music and their positive musical emotions were correlated with left hemisphere activation while their negative emotions were correlated with right hemisphere activation suggesting that a key element in music emotion valence is the syntactic predictability of melody (Schmidt and Trainor, 2001). Thus, the substrate of musical emotion cannot be said to be solely dependent on both direct and indirect (i.e. cognitively mediated) effects on subcortical emotional circuits of the human brain that seem to be essential for generating affective processes as defended by Panksepp and Bernatzky (2002), but dependent on the activation of extensive cortical and subcortical networks that gain their emotional meaning only in higher neocortical substrata intersecting with the language systems of the brain. Perhaps it would be helpful to distinguish *basic emotions*, largely dependent on subcrotical limbic structures, from *higher emotions* partaking with cognitive meaning. One type of such higher emotions may be moral emotions, which depend on belief systems (Mercadillo *et al.,* 2008), and another would be musical emotions, which depend on melodic syntax and sound symbols (Lerdahl and Jackendoff, 1983; 1996) to integrate non-referential music representations (Raffmann, 1993).

It seems feasible to pose that the large neuronal networks involved in the processing of agreeable music and involving the dominant hemisphere for language are to some extent the same elements involved in the processing of language structures. In this way a narrative structure is employed in language and in music for the exposition of feelings using a common basic substratum but from such base two different media. The two different media become apparent because music has not the same characteristics of language such as connotations or literal meanings, but it is likely that it follows some syntactical rules that may be crucial to our liking or disliking particular music pieces. In our experiments this possibility is suggested by the fact that two structurally very different pieces by Bach and Mahler share a wide left hemisphere processing network probably because they fulfil certain syntactic expectations, while the disagreeable piece by Prodromidès, perceived as deprived of syntactic rules, engages the right hemisphere.

These results agree with Diana Raffmann's (1993) idea that there is a connotative relationship between music and subjective emotional experience defined by a certain similarity in logical form, and with Zbikowski (2006) notion that higher cognitive abilities are necessarily implicated in the proper understanding of music. It perhaps could be posed that certain aspects of mind/brain physiology have formal properties similar to the music patterns of motions and rest, tension and release, agreement and disagreement that may be crucial in the emotional connotations of music experience. It seems possible to consider that music patterned processes elicit coherent or incoherent patterns among brain modules and that such level of coherence is interpreted with a related gradient of emotional valence in the sense that coherent patterns are perceived or felt as pleasant and incoherent patterns as unpleasant.

Nevertheless one must also agree that music emotion depends and develops from neuronal circuits at several levels of auditory input and brain processing (Juslin and Västfjäll, 2008) but also that neocortical cognitive mechanisms are involved. Our remarkable result that primary auditory areas are involved in the qualitative appreciation of music concurs with such a claim. The picture emerging from this analysis is that music emotion, very much in accord with other higher mental functions, is progressively integrated at the different relays of a pathway that does not end in primary sensory areas, but that must be processed by inter-modular dynamics in extensive areas of the brain to integrate conscious experience (Díaz, 2007).

Finally, in terms of the nature of musical emotions considered within the frame of the emerging neuroaesthetics paradigm, and in accord with our empirical results, we also agree with the contention that such emotions differ from those evoked by ecological and social stimuli predominantly engaging limbic and other subcortical and paleocortical brain regions.

The astonishing mimetic affective powers of music require higher symbolic mind/brain mechanisms developed and functioning in unison with cultural cosmologies and resulting in peculiar emotional experiences.

All art objects represent artificial stimuli that are implicitly taken by the viewer or listener to be different in kind from natural stimuli in the extended mimetic sense that play objects are taken to be (Walton, 1990). The artificial structures called art forms, such as music "playing," are able to perform, express and carry symbolic meanings and values capable of extending and even transcending natural stimuli by impinging into deeply meaningful human emotions, questions, and values engaging extensive and deeply semantic neocortical neuron networks.

References

[1] Altenmüller, E.O., Shurman, K., Lim, V.K., Parlitz, D., 2002. Hits to the left, flops to the right: different emotions during listening to music are reflected in cortical lateralization patterns. *Neuropsychologia* **40**, 2242–2256.

[2] Andreason, P.J., Zametkin, A.J., Guo, A.C., Baldwin, P., Cohen, R.M., 1994. Gender-related differences in regional cerebral glucose metabolism in normal volunteers. *Psychiatry Research* **51**, 175-83.

[3] Bachelder, L. (1975) *The Gift of Music*. Mount Vernon, New York, The Peter Pauper Press.

[4] Balkwill, L., Thompson, W.F. (1999). A cross-cultural investigation of the perception of emotion in music: psychophysical and cultural cues. *Music Perception* **17**, 43–64.

[5] Baumgartner, T., Lutz, K., Schmidt, C.F., Jäncke, L. (2006). The emotional power of music: how music enhances the feeling of affective pictures. *Brain Research* **1075** (1), 151–164.

[6] Bhattacharya, J., Petsche, H., Pereda, E. (2001). Long-range synchrony in the gamma band: role in music perception. *Journal of Neuroscience* **21**, 6329–6337.

[7] Blood, A.J., Zatorre, R.J. (2001). Intensely pleasurable responses to music correlate with activity in brain regions implicated in reward and emotion. *Proceedings of the National Academy of Sciences of the United States of America* **25**: 98(20); 11818–11823.

[8] Blood, A.J., Zatorre, R.J., Bermudez, P., Evans, A.C. (1999). Emotional responses to pleasant and unpleasant music correlate with activity in paralimbic brain regions. *Nature Neuroscience* **2,** 382–387.

[9] Brown, S. (2001). Are music and language homologues? *Annals of the New York Academy of Sciences* **930**, 372–374.

[10] Brown, S., Martínez, M.J., Parsons, L.M. (2004). Passive music listening spontaneously engages limbic and paralimbic systems. *NeuroReport* **15**, 2033–2037.

[11] Campanella, S., Rossignol, M., Mejias, S., Joassin, F., Maurage, P., Debatisse, D., Bruyer, R., Crommelinck, M., Guérit, J.M. (2004). Human gender differences in an emotional visual oddball task: an event-related potentials study. *Neuroscience Letters* **367**, 14–18.

[12] Christensen, P.G., Peterson, J.B. (1988). Genre and gender in the structure of music preferences. *Communications Research* **15**, 282–301.

[13] Damasio, A.R. (1989). Time-locked multiregional retroactivation: a systems level proposal for the neural substrates of recall and cognition. *Cognition* **33**, 25–62.

[14] Darwin, C. (1872) *The Descent of Man and Selection in Relation to Sex.* New York: Appleton and Co.

[15] Davidson, R.J. (1992). Anterior cerebral asymmetry and the nature of emotion. *Brain and Cognition*. **20**, 125–151.

[16] Davies, S. (1997). Why listen to sad music if it makes one feel sad? In: Jenefer Robinson (ed.) *Music and Meaning.* Ithaca and London: Cornell University Press, pp. 242-253.

[17] Díaz, J.L. (1997) A patterned process approach to brain, consciousness, and behaviour. *Philosophical Psychology* **10**:179-195.

[18] Díaz, J.L., Flores, E. (2001). La estructura de la emoción humana: un modelo cromático del sistema afectivo. *Salud Mental* (México) **24** (4), 20–35.

[19] Díaz, J.L. (2007) *La conciencia viviente.* México, Fondo de Cultura Económica.

[20] Ekman, P. (1992). An argument for basic emotions. *Cognition and Emotion*, **6,** 169-200.

[21] Flores-Gutiérrez, E. O. (2001). *La respuesta emocional a la música: atribución de términos de la emoción a segmentos musicales*. Thesis for Master in Neurobiology. México: Universidad Nacional Autónoma de México.

[22] Flores-Gutiérrez E. O., Díaz J.L., Barrios F.A., Favila-Humara R, Guevara M.A., del Río-Portilla Y., Corsi-Cabrera M. (2007). Metabolic and electric brain patterns during pleasant and unpleasant emotions induced by music masterpieces. *International Journal of Psychophysiology* **65**(1), 69-84.

[23] Flores-Gutiérrez, E. O., Díaz J. L. (2009). The emotional response to music: attribution of emotion words to musical segments. *Salud Mental* (México) **32**(1), 21-34.

[24] Flores-Gutiérrez, E. O., Díaz, J. L., Barrios, F. A., Guevara, M. A., Del Río-Portilla, Y., Corsi-Cabrera, M., Flores-Gutiérrez, E. O. (2009). Differential alpha coherence hemispheric patterns in men and women during pleasant and unpleasant musical emotions. *International Journal of Psychophysiology* **71**(1), 43-49.

[25] Freedberg, D. (2006) Composition and emotion. In: *The Artful Mind. Cognitive Science and the Riddle of Human Creativity.* (Mark Turner, editor) Oxford University Press, pp 73-92.

[26] Fukui, H. (2001). Music and Testosterone: a new hypothesis for the origin and function of music. *Annals of the New York Academy of Sciences*, **930**, 448-451.

[27] Gagnon, L., Peretz, I. (2000). Laterality effects in processing tonal and atonal melodies with affective and non-affective task instructions. *Brain and Cognition*. **43**, 206–210.

[28] Goldstein, J. M., Seidman, L. J., Horton, N. J., Makris, N., Kennedy, D. N., Caviness, V. S., Jr Faraone, S. V., Tsuang, M. T. (2001). Normal sexual dimorphism of the adult human brain assessed by in vivo magnetic resonance imaging. *Cerebral Cortex* **11**, 490-497.

[29] Gur RC, Schroeder L, Turner T, McGrath C, Chan RM, Turetsky BI, Alsop D, Maldjian J, Gur RE. (2002). Brain activation during facial emotion processing. *Neuroimage* **16**(3), 651-662.

[30] Gur, R. C., Mozley, L. H., Mozley, P. D., Resnick, S. M., Karp, J. S., Alavi, A., Arnold, S. E., Gur, R. E. (1995). Sex differences in regional cerebral glucose metabolism during a resting state. *Science* **267**, 528-31.

[31] Hall, J. (1978). *Nonverbal Sex Differences: Communication Accuracy and Expressive Style*. Johns Hopkins University Press, Baltimore.

[32] Halpern, A. R., Zatorre, R. J., Bouffard, M., Johnson, A. (2004). Behavioral and neural correlates of perceived and imagined musical timbre. *Neuropsychologia* **42** (9), 1281–1292.

[33] Herrington, J. D., Mohanty, A., Koven, N. S., Fisher, J. E., Stewart, J. L., Banich, M. T., Webb, A. G., Miller, G. A., Heller, W. (2005). Emotion-modulated performance and activity in left dorsolateral prefrontal cortex. *Emotion* **5** (2), 200–700.

[34] Hevner, K. (1936). Experimental studies of the elements of expression in music. *American Journal of Psychology* **48**, 246–268.

[35] Huron, D. (2001). Is music an evolutionary adaptation? *Annals of the New York Academy of Sciences* **930**, 43–1.

[36] Janata, P., Birk, J. L., vHorn, J. D., Leman, M., Tillman, B., Bharucha, J. J. (2002). The cortical topography of tonal structures underlying western music. *Science* **298**, 2167–2170.

[37] Johnson-Laird, P. N., Oatley, K. (2008) Emotions, music and literature. In: Lewis, M., Haviland-Jones, J., and Feldman-Barrett, L.F. (Eds.). *Handbook of Emotions*, third edition. New York: Guilford Press.

[38] Jolliffe, L. T. (1986). *Principal Component Analysis*. Springer-Verlag, New York.

[39] Juslin, P. N., Sloboda, J. A., eds (2001) *Music and emotion: Theory and research.* Oxford: Oxford University Press.

[40] Juslin, P. N. (2005). From mimesis to catharsis: expression, perception, and induction of emotion in music. In (D. Miell, R. MacDonald, and D. J Hargreaves Eds.), *Musical communication* (pp. 85-115). New York: Oxford University Press.

[41] Juslin, P. N. and Västjäll, D. (2008) Emotional responses to music: The need to consider underlying mechanisms. *Behavioral Brain Sciences* **31**: 559-621.

[42] Koelsch, S., Fritz, T., v Cramon, D.Y., Muller, K., Friederici, A.D. (2006). Investigating emotion with music: an fMRI study. *Human Brain Mapping* **27**, 239–250.

[43] Koelsch, S., Maess, B., Grossmann, T., Friederici, A.D. (2003). Electric brain responses reveal gender differences in music processing. *NeuroReport* **14**, 709–713.

[44] Krumhansl, C.L. (2003). Dissecting the perceptual components of music. *Annals of the New York Academy of Sciences* **999**, 103–105.

[45] Langer, S. K. (1942/1951) *Philosophy in a New Key*. New York: Mentor Books.

[46] Lerdahl, F., Jackendoff, R. (1983/1996) *A Generative Theory of Tonal Music*. Cambridge: MIT Press.

[47] Levinson, J. (1997) *Music and negative emotion*. In: Jenefer Robinson (ed.) *Music and Meaning*. Ithaca and London: Cornell University Press, pp. 215-241.

[48] Mandal, M.K., Palchoudhury, S., (1985). Perceptual skills in decoding facial affect. *Perceptual and Motor Skills*. **60**, 96–98.

[49] Mercadillo, R.E., Díaz, J.L., Barrios, F.A. (2007) Neurobiología de las emociones morales. *Salud Mental (México)* **30** (3): 1-11.

[50] Mesulam, M.M., 1998. From sensation to cognition. *Brain* **121**, 2–52.

[51] Meyer, L., 1956. *Emotion and Meaning in Music*. The University of Chicago Press, Chicago, Ill.

[52] Michel, C. M., Thut, G., Morand, S., Khateb, A., Pegna, A. J., Grave de Peralta, R., Gonzalez, S., Seeck, M., Landis, T. (2001). Electric source imaging of human brain functions. *Brain Research. Brain Research Reviews*. **36**, 108–118.
[53] Newman, J. (1997). Putting the puzzle together, Part I: towards a general theory of the neural correlates of consciousness. *Journal of Consciousness Studies* **4**, 47–66.
[54] Ogata, S. (1995). Human EEG responses to classical music and simulated white noise: effects of a musical loudness component on consciousness. *Perceptual and Motor Skills* **80**, 79–90.
[55] Panksepp, J., Bernatzky, G. (2002) Emotional sounds and the brain: the neuro-affective foundations of musical appreciation. *Behavioral Processes* **60**(2), 133-155.
[56] Patel, A. D., Balaban, E. (2000). Temporal patterns of human cortical activity reflect tone sequence structure. *Nature* **404**, 80–84.
[57] Patterson, R. D., Uppenkamp, S., Johnsrude, I. S., Griffiths, T. (2002). The processing of temporal pitch and melody information in auditory cortex. *Neuron* **36**, 767–776. *Proceedings of the National Academy of Sciences U. S. A.* **98**, 11818–11823.
[58] Pigott, T.A. (1999). Gender differences in the epidemiology and treatment of anxiety disorders. *Journal of Clinical Psychiatry* **60**:Suppl 18, 4-15.
[59] Podell, K., Lovell, M., Goldberg, E. (2001). Lateralization of frontal lobe functions. In: Salloway, S. P., Malloy, P. F., Duffy, J. D. (Eds.), *The Frontal Lobes and Neuropsychiatric Illness.* American Psychiatric Publishing, London, pp. 83–99.
[60] Raffmann, D. (1993) *Language, Music, and Mind.* Cambridge, MA, MIT Press. A Bradford Book.
[61] Ramos, J., Corsi-Cabrera, M. (1989). Does brain electrical activity react to music? *International. Journal of Neuroscience*. **47**, 351–357.
[62] Ramos, J., Guevara, M. A., Martínez, A., Arce, C., Del Rio, Y., Amescua, C., Corsi-Cabrera, M. (1996). Evaluación de los estados afectivos provocados por la música. *Revista Mexicana de Psicología* **13**, 131–145.
[63] Sakai, K., Hikosaka, O., Miyauchi, S., Ryousuke, T., Tamada, T., Iwata, K. N., Nielsen, M. (1999). Neural representation of a rhythm depends on its interval ratio. *Journal of Neuroscience*. **19**, 10074–10081.
[64] Samson, S., Ehrlé, N., Baulac, M. (2001). Cerebral substrates for musical temporal processing. *Annals of the New York Academy of Science*. **930**, 166–178.
[65] Schirmer, A., Kotz, S.A. (2003). ERP evidence for a sex-specific Stroop effect in emotional speech. *Journal of Cognitive Neuroscience*. **15**, 1135–1148.
[66] Schmidt, L. A., Trainor, L. J. (2001). Frontal brain electrical activity (EEG) distinguishes valence and intensity of musical emotions. *Cognition and Emotion*. **15**, 487–500.
[67] Seashore, C. E. (1938/1967) *Psychology of Music*. New York, Dover Publications Inc.
[68] Walton, K. (1990). *Mimesis as Make-Believe: On the Foundations of the Representational Arts*. Cambridge, MA: Harvard University Press.
[69] Wieser, H. G., Mazzola, G. (1986). Musical consonances and dissonances: are they distinguished independently by the right and left hippocampi? *Neuropsychologia* **24**, 805–812.
[70] Zatorre, R., Evans, A., Meyer, E. (1994). Neural mechanisms underlying melodic perception and memory for pitch. *Journal of Neuroscience*. **14**, 1908–1919.

[71] Zatorre, R.J. (2001). Neural specializations for tonal processing. *Annals of the New York Academy of Science*. **930**, 193–210.

[72] Zbikowski, L. B. (2006) The cognitive tango. In: *The Artful Mind. Cognitive Science and the Riddle of Human Creativity*. (Mark Turner, editor) Oxford University Press, pp 115-132.

[73] Zeki, S. (1999/2003) *Inner Vision. An Exploration of Art and the Brain*. Oxford University Press.

In: Music: Composition, Interpretation and Effects
Editor: Tamara E. Ivanova, pp. 57-74
ISBN: 978-1-60876-170-8

Chapter 3

THE IMPACT OF ADOLESCENTS' MUSIC VIDEO VIEWING ON RISKY DRIVING

***Kathleen Beullens**, Keith Roe*• and Jan Van den Bulck*♦**
Leuven School for Mass Communication Research,
Katholieke Universiteit Leuven, Leuven, Belgium

Abstract

Purpose: Music video viewing is a very popular pastime among adolescents. The public criticism of music videos has encouraged researchers to examine the effects of music video exposure. Although an association between music video exposure and several health risk behaviors (e.g. drinking) has been found, the relationship between music video viewing and risky driving has remained largely unexamined. In this study the relationship between adolescents' music video viewing and risky driving such as driving after consuming alcohol and joy riding is explored.
Methods: Participants were 354 adolescent males and females who participated in a panel study (2-year interval). Respondents were 17 or 18 years old during the first wave of data collection and did not have their driver's license yet. They completed a questionnaire on music video viewing, sensation seeking, aggression, attitudes towards joy riding and driving after the consumption of alcohol, and the intention to perform these behaviors in the future. Two years later the respondents had obtained their driver's license and were questioned on their actual risky driving behavior. The relationships between these constructs were analyzed using structural equation models.
Results: The results indicate that music video viewing is indirectly associated with joy riding and driving after the consumption of alcohol through the attitudes towards these behaviors and the intention to perform these behaviors in the future. More music video viewing resulted in a more positive attitude towards risky driving, even after controlling for sensation seeking and aggression. These attitudes are positively related to the intention to perform these reckless behaviors in the future and these intentions are, in turn, a good predictor of the actual risky driving behavior two years later.

* E-mail address: Kathleen.Beullens@soc.kuleuven.be. Tel: +32(0)16.32.32.19// +32(0)16.32.32.20 Fax: +32(0)16.32.33.12. Corresponding author: Kathleen Beullens, Parkstraat 45 (box 3603), 3000 Leuven (BELGIUM).
• E-mail address: Keith.Roe@soc.kuleuven.be.
♦ E-mail address: Jan.VanDenBulck@soc.kuleuven.be.

Conclusions: The results indicate that music video viewing during adolescence is an important and significant predictor of risky driving two years later.

Introduction

Music video viewing is a very popular pastime among adolescents. A recent study by Beentjes, Konig and Krzeszewski (2008) found that adolescent boys as well as girls watch rap/hip-hop music videos at least 27 minutes a day on average. Several other studies have reported frequent music video viewing in different samples of young people (Sherman and Dominick, 1986; Sun and Lull, 1986). Van den Bulck and Beullens (2005) found that 40.8% of the 16-year-old girls and 36.2% of the boys of the same age viewed music videos almost every day. Another study by the same authors indicated that watching music videos was also a popular activity for slightly older adolescents (Beullens and Van den Bulck, 2008).

Sun and Lull (1986) and Hansen and Hansen (2000) ascribe the popularity of music videos, first, to the positive effects of contemporary music on the mind and body of its listeners. Music listening induces generally pleasant feelings and reduces anxieties. Second, peer groups are often formed around specific musical preferences. As such music listening is important for the construction of both a self–identity and a group-identity. Third, they see music as an "agent of socialization". Through the content of popular music young people learn the unwritten rules of their peers, which facilitates in-group identification and communication. By providing role models for adolescents, music artists also play a role in the identity formation of adolescents.

The popularity of music videos has encouraged researchers to examine their content and their possible effects. Content analyses have indicated that there may be cause for concern. Gruber et al. (2005) found frequent references to the use of illicit substances, alcohol, and tobacco in their sample of music videos. 43% of the videos contained at least some evidence of these health risks. They appeared to be more prevalent in rap and hip-hop songs compared to other music videos. These results were in line with DuRant et al.'s (1997b), who reported that rap music videos depict the most instances of smoking (30% of the videos in this genre) and alcohol use (27%), compared to other genres. Nevertheless health risk behaviors such as smoking and drinking were frequently encountered in other music genres also (DuRant et al, 1997b; Baxter et al., 1985).

Risky sexual behavior also appears to be portrayed in music videos regularly. Baxter et al. (1985) stated that music television's content is highly sexual. 60% of the music videos in their sample contained images of sexual feelings or impulses such as embrace or other physical contact, provocative clothing, movements of a sexually suggestive nature etc. These results were confirmed by Sherman and Dominick (1986). Their study revealed that more than 75% of the music videos presented visibly sexual intimacy. These depictions were mostly implied rather than overt and were observed among different genres of music videos (Tapper and Thorson, 1994).

Music videos' violent content has also received a lot of attention (e.g. Sherman and Dominick, 1986; Baxter et al., 1985; Sommerflanagan, Sommersflanagan and Davis, 1993). DuRant et al. (1997a) found that a substantial number of music videos portrayed violence and weapon carrying. About 20% of rap and rock videos contained overt violence. These depictions were found to be glamorized by the performers. According to Rich et al. (1998)

15% of the videos depicted violent behavior. Perpetrators of these acts were mostly males (78%). Similarly Smith and Boyson (2002) noted that 15% of music videos featured physical aggression. This appeared to be presented in an often realistic manner and without showing any negative consequences for the aggressors.

Following DuRant et al. (1997a, 1997b) it may be argued that given the frequent depictions of health risk behaviors in music videos, even moderate viewers are frequently exposed to it. As several authors have stated, these positive portrayals of risky behaviors may influence their viewers (Bryant and Zillmann, 2002). Cultivation Theory proposes that frequent exposure to similar television content may, in the long term, lead to alterations in its viewers' perceptions and attitudes (Gerbner and Gross, 1976; Gerbner, Gross, Morgan and Signorielli, 1986). Social Cognitive Theory also suggests that exposure to music television may influence adolescents, but explains this relationship by the fact that popular artists, who serve as role models for adolescents, are engaging in the risky behaviors in music videos (Bandura, 2002). In these portrayals the engagement in risky behaviors is often legitimized, trivialized and glamorized thus potentially altering adolescents' perceptions of what is socially sanctioned (Bandura, 2002). As such it may reduce inhibitions to perform a behavior and may desensitize viewers. Ultimately this may lead to participation in the depicted forms of risk taking (Bandura, 2002; DuRant et al., 1997b). Priming theory argues that the exposure to certain images on television may trigger thoughts that are semantically related and influence later behavior (Roskos-Ewoldson, Roskos-Ewoldson and Carpentier, 2002).

Following these media-effects theories, researchers have examined the relationship between music video viewing and several risky behaviors. Beentjes et al. (2008) found the amount of rap/hip-hop video viewing to be a significant predictor of the idea (among men and women) that women behave as sex objects and of the belief (only among men) that men are the most dominant partner in sexual relationships. Zhang, Miller and Harrison's (2008) study also indicated a relationship between exposure to music videos and sexual attitudes. The viewing of music videos with a sexually explicit content appeared to be positively related to participants' approval of premarital sex and their score on the double sexual standard scale. Antisocial behavior has also been shown to be associated with the consumption of music videos. Hansen and Hansen's (1990) experiment showed that even brief exposure to music videos with antisocial behavior such as a joyriding or defying parents and police resulted in more favourable evaluations of people engaging in antisocial behavior. Likewise it has been found that greater exposure to rap videos is correlated with the chance to have hit a teacher, to have been arrested, to have had a lot of sexual partners and to have used drugs or alcohol 12 months later (Wingood et al., 2003).

Although several authors have examined the possible effects of music videos, the relationship with risky driving has remained largely unexamined.

Aim of the Study

The aim of the current study is to examine whether frequent exposure to music videos is associated with several forms of risky driving. Statistics show that traffic accidents remain a cause for concern in many industrialised countries (World Health Organization, 2004). Young men appear to be at greater risk of being involved in a motor vehicle accident than most other groups in society (Belgische Instituut voor Verkeersveiligheid (BIVV), n.d., p. 16; Harré,

2000, p. 206). This is only partly explained by their inexperience as this group appears to be more inclined to take risks than other drivers (Arnett, Irwin and Halpern-Felsher, 2002; Ulleberg and Rundmo, 2002). Although it has been argued that there is a need for the examination of the portrayal of driving in the media and its effects (Arnett et al. 2002; Harré, 2000; Reinhardt-Rutland, 2007), virtually no research has dealt with this topic. This is even more remarkable given the fact that traffic accidents are an important cause of injury and death, and that other problem behavior has been found to be associated with media use. Therefore, the main research question of the current study is:

> RQ: Is the viewing of music videos during adolescence associated with risky driving two years later?

This research question draws, upon a combination of cultivation theory (Gerbner and Gross, 1976; Gerbner et al., 1986) and Theory of Planned behavior (Ajzen, 1991; Fishbein and Ajzen, 1975). According to cultivation theory heavy viewers are constantly exposed to similar images on television. In the long term this exposure may result in a world view that is similar to the one that is depicted on television. Consequently the frequent exposure to images of risky driving on television may result in the development of perceptions and attitudes in line with these portrayals in heavy television viewers (Gerbner and Gross, 1976; Gerbner et al., 1986).

Several content analyses have indicated that risky driving is very prevalent on television and that the risks associated with it are hardly ever shown. For Instance Greenberg and Atkin (1983) found that risky driving occurs more than 7 times an hour. Within a sample of 784 driving scenes they observed 1301 risky driving acts. Within these acts quick breaking, squealing brakes and screeching tires occurred most commonly. These risky driving acts appeared to be followed by death, injuries or legal penalties only seldom. Physical damage such as broken cars was slightly more prevalent. Furthermore the rate of the depiction of seatbelt use among drivers appeared to be rather low. Only 25% of the drivers were shown while wearing a seatbelt. Similarly, based on an analysis of the portrayal of risk behavior on prime –time television Will, Porter, Geller and DePasquale (2005) concluded that 26% of the television drivers wore seat belts. In 20% of the scenes risky driving such as speeding or neglecting a stop sign was observed. The potentially negative consequences associated with these behaviors were portrayed only infrequently. In 94% of the scenes there were no positive or negative consequences, the remaining 6% of the scenes comprised car crashes, wounded persons, police intervention and only one fatality. Other content analyses of the portrayal of risk taking on television (Greenberg and Gregg, 1998), movies (Greenberg and Thanki, 1997; Jacobson et al., 2001; Pelletier et al., 2000), reality television (Cowan, Jones and Ho, 2006), and even children's programming (Glik et al., 2005; Winston et al., 2000) came to similar conclusions. They indicated that, regardless of the prevalence of risk taking in television content, the harmful or beneficial consequences of this behavior were rarely shown and preventive practices such as seat belt wearing were hardly ever portrayed.

Unfortunately only one content analysis of the depiction of risky driving in music videos was found. The early study of Baxter et al. (1985) indicated that transportation defined as the use of various types of vehicles or modes of conveyance, occurred in 35.5% of the music videos in their sample. How driving was framed in music videos was not examined in this study. However, given the fact that risky driving is often depicted in other program genres and that other risk behavior is frequently depicted in music videos, it is assumed that risky driving is very prevalent in music videos as well. Following this assumption and cultivation theory it can be hypothesized that the frequent viewing of music videos is associated with a more positive attitude towards risky driving.

Following the Theory of Planned behavior it can be assumed that these attitudes are an important indirect predictor of behavior, through the intention to perform this behavior (Ajzen, 1991; Perloff, 2003, p. 90). Behavioral intentions are assumed to "... capture the motivational factors that influence a behavior, they are indicators of how hard people are willing to try, of how much effort they are planning to exert, in order to perform the behavior." (Ajzen, 1991, p. 181). Within this framework, intention is held to be partly determined by attitude, that is, a general positive or negative evaluation of the behavior. A meta-analysis of 185 TPB studies addressing health- and risk related behaviors showed that intention was, in fact, the strongest predictor of subsequent behavior, and that attitudes appeared as the strongest predictor of behavioral intentions (Armitage and Conner, 2001).

Thus, although cultivation theory only deals with changes in perceptions or attitudes as a result of exposure to media messages and therefore does not in itself explain changes in behavior, social psychological theories such as the Theory of Planned Behavior have looked extensively at the relationship between attitudes and behavior. The theoretical framework used in this study is based on the combination of both theories. It is assumed that the content of music videos is responsible for a possible relationship between music video viewing and risk behavior and that long-term exposure to music video viewing leads to changes in attitudes and ultimately behavior. This resulted in the following hypotheses.

- H1: The amount of music video viewing is a positive predictor of the attitudes towards risky driving.
- H2: Attitudes towards risky driving are a good positive predictor of the intention to engage in this behavior.
- H3: The intention to engage in risky driving during adolescence is a positive predictor of actual driving behavior two years later.

The model proposed in this chapter controlled for two personality characteristics: sensation seeking and aggression. An extensive amount of research has shown that these characteristics are associated with both risky driving and media use (a.o. Jonah, 1997; Arnett, Offer and Fine, 1997; Chliaoutakis, Demakakos et al., 2002; Dahlen and White, 2006;Ulleberg and Rundmo, 2003; Sümer, 2003). Arnett (1996), for instance, found that aggression and sensation seeking were positively associated with several forms of risk taking in traffic. Jonah's (1997) literature review indicated that a large majority of studies posited that sensation seeking is a positive predictor of several forms of reckless driving behavior. Correlations ranged between 0.30 and 0.40 depending on the scales used and the subgroups under examination. Gulliver and Begg (2007) found in their research on the personality factors predicting persistent risk taking in traffic that aggression was associated with risky driving behavior and crash risk. Ulleberg and Rundmo (2003) also emphasized the importance of the inclusion of personality characteristics in models examining the predictors of risky driving. They, however, propose a model in which sensation seeking and aggression (and a number of other characteristics) influence several forms of risky driving such as speeding and fun riding indirectly through the attitudes towards these behaviors. Fishbein (2000) and Fishbein and Cappella (2006) similarly regarded personality characteristics as indirect predictors of health behavior. Fishbein (2000) argued that personality characteristics predict perceptions. These in turn are the predictors of attitudes, which are, among others, the predictors of intentions. (Fishbein and Cappella, 2006).

Research has also indicated, however, that media use is associated with several personality characteristics (Perse, 1996; McNamara and Ballard, 1999; Potts, Dedmon and Halford, 1996; Weisskirch and Murphy, 2004; see Hall (2005a; 2005b) for an overview). People high in sensation seeking, for instance, appear to prefer arousing media content such as action and horror movies or erotic programs, and are more frequently exposed to these kind of programs (Perse, 1996; Zuckerman and Litle, 1986; Edwards, 1991; Schierman and Rowland, 1985). Low sensation seekers on the other hand seem to avoid violent or sexual media content and have a preference for programs such as musical films and romantic stories. These genres are in turn not appealing to viewers high in sensation seeking. Aggression has also been associated with a preference for specific media content. Aluja-Fabregat (2000), for example, found that adolescent boys who watch a lot of violent movies are more frequently described as aggressive and short-tempered by their teachers. Given the fact that music videos are expected to be rather arousing media and that research has shown that people high in sensation seeking and aggression prefer arousing media content, it is hypothesized that:

- H4: Aggression and sensation seeking are (direct) positive predictors of music video viewing.
- H5: Aggression and sensation seeking predict risky driving directly and indirectly through media use and the attitudes towards risk taking.

The analyses controlled for total amount of television viewing as this study wants to examine the specific effect of music television viewing.

In this chapter, two specific forms of risky driving are under examination, namely fun riding or joy riding and driving after the consumption of alcohol. It is assumed that both examples of risky driving regularly occur in music videos and therefore, following cultivation theory and TPB, exert an influence on adolescents' attitudes, intentions and behavior. Together hypotheses 1 to 5 lead to the hypothetical model proposed in figure 1.

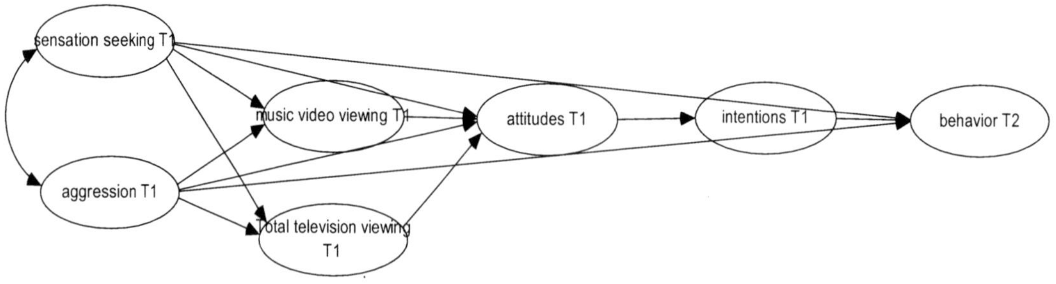

Figure 1. Hypothetical model of the relationship between music video viewing and risky driving behavior.

In the subsequent analyses, this model will be tested for several subgroups. Ajzen (1991, p. 188) suggested that the relative importance of the different concepts included in the TPB may vary across different populations (e.g. boys and girls). Research has also shown that, although all risk groups underestimate their actual crash risks, high risk groups such as young men under-assess their personal road-traffic risk even more than other risk groups do (e.g. young women) (Andersson and Lundborg, 2007). Therefore we expect gender to be a significant moderator of the relationship between music television viewing and risky driving.

In sum, in the current study, a two wave panel study was used to examine whether adolescent boys' and girls' music video viewing predicts their risky driving two years later. Following cultivation theory and TPB it is hypothesized that this relationship is mediated through the attitudes towards reckless driving behavior and the intention to perform this behavior in the future.

Method

Sample

For the first wave of data collection questionnaires were administered to a stratified random sample of secondary school students in Flanders, Belgium. From the official list of secondary schools in Flanders 20 schools were randomly selected. These schools were contacted with the request to cooperate in a large-scale study on the relationship between media use and risk behavior among adolescents. When a school agreed to cooperate, all students from the fifth and sixth year were included in the sample. This selection procedure was repeated until 15 schools agreed to cooperate in the study.

In the weeks following sample selection research assistants visited the selected schools to administer the questionnaires. The study was presented to the pupils as a study on leisure activities (part 1) and traffic (part 2). In total 2193 pupils filled out a standardized, self-administered questionnaire including measures of television viewing and risky driving. 65.2% of the students were boys, 34.8% were girls. 41.7% of the respondents were born in 1988, 36.5% in 1989 (M=1988, StD=.93), thus the large majority of our respondents were 17 or 18 years old at time 1. In Flanders a definitive driver's license can only be obtained at the age of 18. A large majority (91.3%) of the respondents in the first wave did not yet have their driver's license during data collection. As a result, their actual driving behavior was not assessed in the questionnaire.

The strict confidentiality of the respondents' answers was stressed before as well as after the completion of the questionnaire. Next, the respondents were informed that a follow-up study was planned two years later and they were asked whether they agreed to cooperate in the second part of the study. Students who promised to collaborate were asked to provide their home address as well as their e-mail address. In total 89.5% agreed to take part in the second wave of the study and provided their home address, 71.3% provided their e-mail address.

Two years after the first wave of datacollection the second wave was administered. First, a follow-up websurvey was sent to all the students who had provided their e-mail account. 31.5% of the e-mails returned to the sender because of mistakes in the e-mail addresses or because the addresses did not exist anymore. In line with the Total Design Method (Billiet and Carton, 2001), two follow-up e-mails were sent after the first contact. In sum this resulted in the completion of 552 websurveys, this is 51.5% of the respondents who provided a correct e-mail address.

Second, respondents who had not yet answered were sent a paper version of the questionnaire and a stamped and addressed envelope at their home address. Two weeks later a letter was sent to thank all the respondents who had already taken part in the second wave of the study and to encourage the others to fill out the questionnaire and send it back to the researcher. Two weeks later a third reminder (which included a new questionnaire and return envelope) was sent to those respondents who had not yet answered. In total this resulted in 552 completed questionnaires. The postal survey and the websurvey together resulted in a response of 1104 respondents. Thus 50% of the respondents who completed the first questionnaire took part in the second wave of the study.

The aim of the current chapter is to examine the relationship between music video viewing during adolescence and risky driving two years later. Only those respondents who completed both waves and had obtained their driver's license in the second wave of the study were kept for the analyses. This is a subsample of 354 young persons, 62.7% were boys, 37.3% girls, 53.8% were born in 1988, 29.7% in 1989 and 14.4% in 1987. Consequently most of the respondents in the subsample on which the analyses were run were 17 or 18 years old at time 1 and 19 or 20 years old at time 2. 42% of the boys and 31.1% of the girls without driver's license at time 1 had obtained their driver's license at time 2 (χ^2=11.832, df=1, p<.01).

The study and sampling method were approved by the Institutional Ethics Committee of the Katholieke Universiteit Leuven and permission to interview the children was obtained from the legal guardians of the children.

Measures

Music video exposure was measured as part of a long list of television content types. Respondents had to answer 'how often do you watch music videos' on a scale (0) never, (1) a few times a year, (2) about once a month, (3) a few times a month, (4) about once a week, (5) a few times a week, (6) (almost) every day (Van den Bulck, Beullens and Mulder, 2006).

Total television viewing volume was assessed using a timeline for each day of the week. The timelines began at 7.00 A.M. and ended at 04.00 A.M. They consisted of 42 checkboxes, each representing a half hour of possible television viewing time. For each day of the week, respondents were asked to mark the period(s) of time on which they normally watch TV on that particular day. Total television viewing time (in hours) per week was obtained by counting the marked checkboxes for each respondent and dividing the result by two. Total television viewing in hours per day was obtained by dividing that variable by seven (Eggermont, 2006).

Ulleberg and Rundmo (2002) validated an instrument to measure young drivers' risk taking attitudes. This instrument consists of 11 dimensions, 2 of them - attitudes towards joyriding (fun riding) and drinking and driving - were translated into Dutch and used in the questionnaire. A principal components factor analysis with Varimax rotation yielded the same factor structure as Ulleberg and Rundmo found. Cronbach's Alpha was calculated for each factor: funriding (α=. 78) and drinking and driving (α=.75).

Sensation Seeking was questioned using Arnett's Inventory of Sensation Seeking (AISS). Haynes, Miles and Clements (2000) analyzed AISS and Zuckerman's sensation seeking scale Form V and concluded that a modified version of AISS provides a more appropriate measure of sensation seeking. The original AISS consists of 20 items and is made up of two subscales: intensity and novelty. For the current study only the intensity factor was used in the analyses, since it was felt that this factor was most closely related to music video viewing. Principal components analysis indicated that 6 items of the original scale loaded on the intensity factor. Two items were removed from the factor since they referred to media content (e.g. 'I like a movie where there are a lot of explosions and car chases'). 4 variables (items 8, 16, 18, 20 from the original scale) loaded onto the intensity factor and were kept in the analyses (Cronbach's Alpha =. 57). Physical aggression was measured using Buss and Perry's (19) Aggression questionnaire (Cronbach's Alpha=.81). Responses for both personality characteristics ranged from (1) do not agree at all, (2) do not agree, (3) agree to (4) totally agree.

Respondents' intention to engage in fun riding and driving after consuming alcohol was measured as part of a series of questions regarding risky driving behavior. *Intention* has been defined by Fishbein and Ajzen (1975, p. 288) as "[...] a person's subjective probability that he will perform some behavior". Respondents had to indicate whether they thought they would (1) drive faster than allowed, (2) drive while they might have had too much alcohol, (3) drive while they had definitely had too much alcohol, (4) take risks to make driving more fun in the future. Response categories were (0) never, (1) seldom, (2) sometimes, (3) often, (4) very often and (5) always. Item 1 and 4 were used as dependent measure for fun riding. Items 2 and 3 both loaded on the latent drinking and driving variable.

Actual fun riding was measured using 2 questions: (1) How often do you drive fast for the thrill of it?, (2) How often do you take risks when driving to make driving more fun? Response categories were (0) never, (1) seldom, (2) sometimes, (3) often, (4) very often and (5) always. These items are part of Begg and Langley's (2004) risky driving scale.

Driving after the consumption of alcohol was measured by 2 items. In the last year, how often have you (1) driven a car while you may had too much alcohol?, (2) driven a car while you definitively had too much alcohol?

It is the aim of the study to test whether music video viewing succeeds in predicting emerging adults' fun riding and driving after the consumption of alcohol. Therefore music video viewing, attitudes and intentions were measured at time 1 and actual risky driving behavior was measured at time 2.

Analyses

To assess the relationships between music video viewing and fun riding (model 1) and driving after consuming alcohol (model 2), a structural equation model was estimated using Amos ^{TM} 15.0. In the current chapter we are mainly interested in the relationship between music video viewing, risk taking attitudes, risky driving intentions and actual reckless driving. Measures of intensity and aggression were included in the models as potential confounders as it has been shown that these constructs are related to risky driving (Jonah, 1997; Gulliver and Begg, 2004) and media use (Hall, 2005ab). The total amount of television viewing was also added as a control variable. Several other variables have been shown to be explain specific behavior (cfr. Theory of Planned Behavior (Ajzen, 1991)). These constructs were not added to our models since the inclusion of constructs that have been shown to correlate with our dependent variable, but were not hypothesized to correlate with our exposure variables, would have increased the fit of the structural equation models artificially.

Results

Descriptive Statistics

Prevalence of Music Video Viewing

31.6% of the respondents in the sample never viewed music videos, 26.2% watched them several times a year, 13.7% about once a month, 12.8% a few times a month, 10.5% about once a week, 5.2% at least several times a week.

Prevalence of Risky Driving

More than 50% of the respondents in the sample admitted to driving, at least occasionally, faster than allowed because they found this behavior exciting. 14.1% reported doing this sometimes, 5.2% often and 3.2% very often or always.

64.5% reported never taking risks in order to make driving more fun. Nevertheless, 22.8% indicated to engage in this behavior seldom, 9.0% sometimes and 3.8% at least often.

Driving after the consumption of alcohol was less prevalent. 80.6% of the respondents never drove after they might have had too much alcohol, 14.6% admitted to having engaged in this behavior in the previous 12 months seldom, 2% about once a month, 2% a few times a month and 0.9% at least once a week. 91.6% never took the car when they definitely had consumed too much alcohol.

Model 1. Music Video Viewing and Fun Riding

It was hypothesized that the personality characteristics aggression and the intensity factor would directly predict media use, attitudes towards risky driving and actual risky driving behavior. Music television viewing was hypothesized to be a positive predictor of risk taking attitudes, which in turn were expected to predict intentions toward risk taking in traffic. Following TPB intentions were hypothesized to be a good direct predictor of risky driving at time 2. Total amount of television viewing was added to the model as a control variable.

Maximum likelihood estimates provided support for the proposed model (χ^2=384.062; p=.000; df=198; CMIN/df=1.940; CFI=.913; RMSEA=.052). However several of the standardized regression coefficients were not significant (e.g. total television viewing was not a significant predictor of attitudes toward fun riding). Following Aish and Jöreskog (1990, p. 441) a new model was tested without these relationships. The respecified model fitted the data well (χ^2=396.539; p=.000; df=205; CMIN/df=1.934; CFI=.910; RMSEA=.051) (see figure 2). Music video viewing during adolescence appeared to be an indirect predictor of fun riding two years later (standardized regression coefficient γ=.24), while the total amount of television viewing was not. This relationship was mediated through the attitudes towards fun riding and adolescent's intentions to engage in this behavior in the future. In total 36.1% of the variance in fun riding was explained by its predictors in the model.

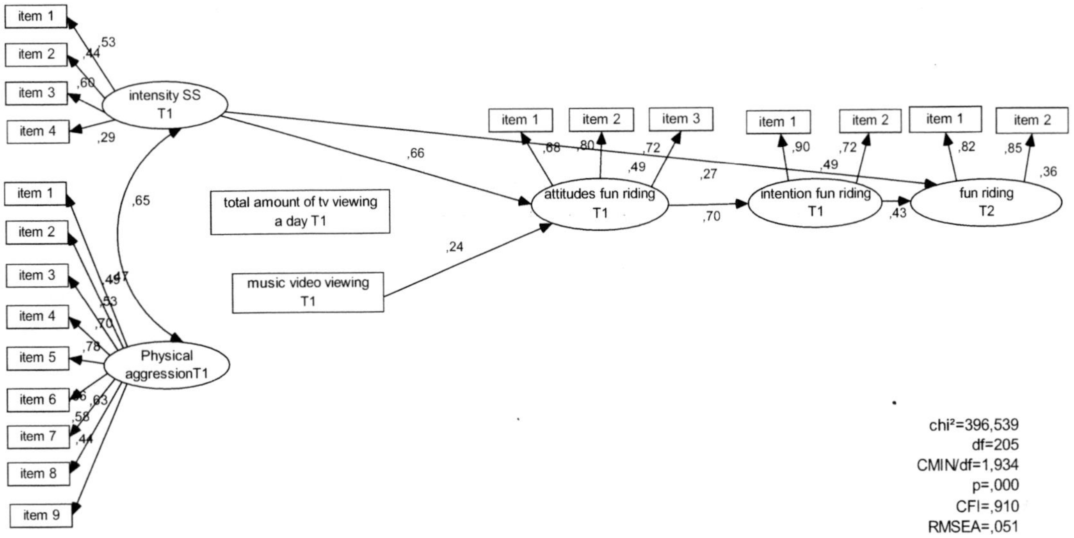

Standardized regressioncoefficients are significant at p<.05.

Figure 2. Model 1: The relationship between music television viewing and fun riding.

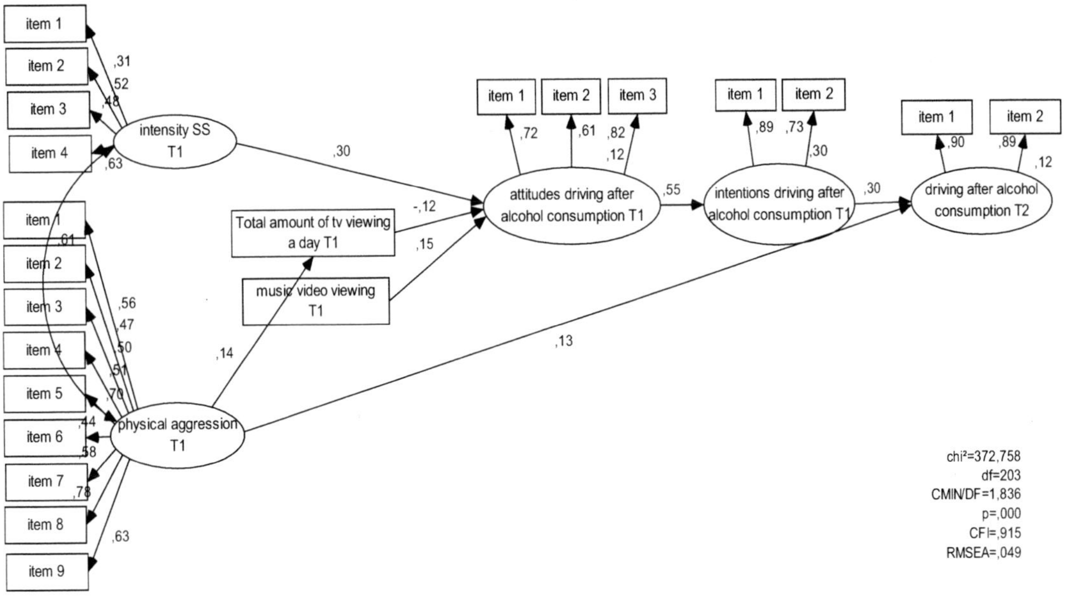

Standardized regression coefficients are significant at p<.05.

Figure 3. Model 2: The relationship between music television viewing and driving after the consumption of alcohol.

A multiple group analysis was executed in order to examine whether gender moderated the relationship between exposure to music television viewing and fun riding. The fit indices indicated that this model fitted the data for boys as well as girls (χ^2=626.321; p=.000; df=410; CMIN/df=1.528; CFI=.871; RMSEA=.039) and that there was no significant difference between the standardized regression coefficients of the relationship between music video viewing and attitudes towards risky driving in both groups.

Model 2. Music Video Viewing and Driving after the Consumption of Alcohol

The second model examined whether music video viewing during adolescence succeeded in predicting a significant part of the variance in driving after the consumption of alcohol two years later. The data provided a good fit for this model (χ^2=370.226; p=.000; df=198; CMIN/df=1.870; CFI=.913; RMSEA=.050). Once more certain relationships appeared to be non-significant and these associations were removed from the model (cf. Aish and Jöreskog, 1990, p. 441). In the renewed version (χ^2=372.758; p=.000; df=203; CMIN/df=1.836; CFI=.915; RMSEA=.049) there was no relationship between the intensity factor of sensation seeking and both measures of television viewing. Aggression was only a significant predictor of total television viewing and was indirectly associated with driving after the consumption of alcohol. As hypothesized, music television viewing was a positive direct predictor of the attitudes toward driving after consuming alcohol (standardized regression coefficient γ=.15). This relationship remained significant when the total amount of television viewing was controlled for. This last variable was negatively related to attitudes (γ=-.12). Music television viewing appeared to be a significantly better predictor of driving after the consumption of alcohol than total television viewing was (Critical ratio for differences =-3.220). Following TPB attitudes were a good predictor of intentions measured at time 1 (γ=.55), intentions (γ=.30) in turn succeeded in predicting a significant part of the variance in actual driving after the consumption of alcohol at time 2. In total 12% of the variance in the dependent variable in the model was explained by its predictors.

Multiple group analyses were used in order to assess whether the model applied to boys as well as girls. Fit indices indicated that the model fitted for both groups (χ^2=592.069; p=.000; df=406; CMIN/df=1.458; CFI=.890; RMSEA=.036). However, an inspection of the standardized regression coefficients in both models indicated that the relationship between music television viewing and attitudes towards driving under the influence of alcohol was significant for girls (γ=.24) but not for boys.

Conclusion

Music video viewing is a very popular pastime among adolescents. Popular debate about the potential impact of music videos has encouraged researchers to examine the effects of music video exposure. Nevertheless, the relationship between music video viewing and risky driving has remained largely unexamined. In this chapter the relationship between adolescents' music video viewing and risky driving such as driving after consuming alcohol and joy riding was explored.

The analyses indicated that adolescents' music video viewing was a small, indirect predictor of their joy riding and driving under the influence of alcohol 2 years later. More music video viewing resulted in a more positive attitude towards risky driving (hypothesis 1). These attitudes predicted adolescents' intentions to drive riskily (hypothesis 2) which in turn predicted their actual later driving behavior (hypothesis 3). This relationship between music television viewing and reckless driving remained significant after controlling for the personality characteristics physical aggression and the intensity factor of sensation seeking and after controlling for the total amount of television viewing per day. This signifies, first, that there is something specific about music videos which cannot be explained by general television viewing. Second, the relationship between media use and risk taking cannot be explained by the personality characteristics that frequent music television viewers and risk takers may both have in common since these characteristics were controlled for in the models. However respecifications had to be made to the models because intensity and aggression did not appear to be good predictors of music television viewing or the attitudes toward risky driving as we had hypothesised. Thus, hypothesis 4 and 5 were only partly confirmed by the data. Multiple group analyses indicated that the models applied to boys as well as girls.

Several explanations can be put forward in order to explain the relationships in our models. First, following cultivation theory and TPB long term exposure to images of risky driving behavior may have an impact on respondents' later driving behavior through a change in attitudes and intentions. We hypothesized that this relationship is explained by the specific content of music videos. We assume that risky driving is very prevalent in music videos and that this behavior is depicted in a glamorized way. Driving behavior in this television genre has, however, not yet been examined fully.

A second explanation for the results found in this study might therefore be that music television viewing and risky driving are both characteristics of a similar lifestyle. Problem Behavior Theory proposes that different kinds of risk taking co-occur and comprise a pattern or cluster of risk-taking (Donovan and Jessor, 1985; Bina et al., 2006). Studies have shown that the viewing of one kind of risk behavior on television may result in engaging in other kinds of risky behavior if these behaviors share a similar meaning for the adolescent (Krcmar and Greene, 2000). Thus, according to PBT, even if no risky driving is depicted in music videos, it is possible that frequent music video viewing and risky driving are both part of a cluster of problem behaviors. In this explanation music television viewing is not the cause of risk taking but another form of risky behavior. This offers a potential explanation of the process by which the viewing of risk behavior in music videos may result in traffic related risk taking. Further research is necessary to determine by which mechanisms the relationship between music television viewing and risky driving occurs and whether music video viewing can be regarded as a cause or as a form of problem behavior.

Despite this discussion about causality, the results of this study remain interesting in that they show that adolescents' music video viewing succeeds in partly predicting their risk behavior two years later. It shows that a group of risky drivers can be identified before they have their driver's license and before they are exposed to the risks. This means that these young people may be addressed (through their media use) with prevention campaigns before they actually start driving and become exposed to the risk of getting hurt in a crash.

The present study has several limitations. First, even though structural equations were used, panel data do not establish causality unequivocally. Experimental studies may be required to examine further the relationship between music television viewing and risky driving. Second, content analytic research should try to asses the portrayal of traffic related risk taking in music videos. If we know whether or not risky driving is present in music videos and how it is framed, this will provide insight into the relationship between these concepts. The current study proposed a theoretical framework which included elements of Cultivation Theory and TPB. However, not all concepts of TPB were assessed in the study and preferably all concepts should be tested by using several measures. Further research should try to integrate these theories in order to explore the processes through which media exposure is related to risky driving (Beullens and Van den Bulck, 2008). For the same reason research should examine the meanings adolescents attribute to specific risk behaviors.

Acknowledgment

This study was funded by a subsidy from the Flemish Fund for Scientific research (FWO).

References

Aish, A.M., and Jöreskog, K.G. (1990). A Panel Model for Political Efficacy and Responsiveness: An Application of LISREL 7 with Weighted Least Squares. *Quality and Quantity,* **24**, 405-426.

Ajzen, I. (1991). The Theory of Planned Behavior. *Organizational Behavior and Human Decision Processes*, **50**, 179-211.

Aluja-Fabregat, A. (2000). Personality and curiosity about TV and films violence adolescents. *Personality and Individual Differences*, **29**(2), 379-392.

Andersson, H., and Lundborg, P. (2007). Perception of own death risk. *Journal of risk and Uncertainty*, **34**(1), 67-84.

Armitage, C.J., and Conner, M. (2001). Efficacy of the Theory of Planned Behaviour: A meta-analytic review. *British Journal of Social Psychology*, **40**, 471-499.

Arnett, J.J. (1996). Sensation seeking, aggressiveness, and adolescent reckless behavior. *Personality and Individual Differences*, **20**(6), 693-702.

Arnett, J.J., Irwin, C.E., and Halpern-Felsher, B.L. (2002). Developmental sources of crash risk in young drivers. *Injury Prevention,* **8**(Suppl. 2), 17-21.

Arnett, J.J., Offer, D., and Fine, M.A. (1997). Reckless driving in adolescence: 'State' and 'trait' factors. *Accident Analysis and Prevention,* **29**(1), 57-63.

Arnett, JJ. (1994). Sensation seeking: A new conceptualization and a new scale. *Personality and Individual Differences,***16**, 289-296.

Bandura, A. (2002). Social cognitive theory of mass communication. In J. Bryant and D. Zillmann (Eds.) *Media effects: Advances in Theory and Research* (2nd Edition). Hillsdale, NJ: Lawrence Erlbaum Associates, 121-154.

Baxter, R.L., De Riemer, C., Landini, A., Leslie, L., and Singletary, W. (1985). A content Analysis of Music Videos. *Journal of Broadcasting and Electronic Media*, **29**(3), 333-340.

Beentjes, H. Konig, R., and Krzeszewski (2008). Muziekvideo's en seksuele opvattingen: Een enquête onder jongeren. *Tijdschrift voor Communicatiewetenschap*, **36**(4), 234-252.

Begg, D. and Langley, J. (2004). Identifying predictors of persistent non-alcohol or drug-related risky driving behaviours among a cohort of Young adults. *Accident Analysis and Prevention,* **36**(6), 1067-1071.

Belgisch Instituut voor de Verkeersveiligheid BIVV, (n.d.). *BIVV verkeersveiligheid Statistieken 2001*. Retrieved September 10, 2006, from http://www.bivv.be/

Beullens, K., and Van den Bulck, J. (2008). News, music videos and action movie exposure and adolescents' intentions to take risks in traffic. *Accident Analysis and Prevention* **40**, 349 356.

Billiet, J., and Carton, A. (2001). Dataverzameling: gestandaardiseerde interviews en zelf-in-te vullen vragenlijsten. In J. Billiet and H. Waege (Eds.) *Een samenleving onderzocht. Methoden van social-wetenschappelijk onderzoek.* Antwerpen, Standaard Uitgeverij, nv.

Bina, M., Graziano, F., and Bonino, S. (2006). Risky driving and lifestyles in adolescence. *Accident Analysis and Prevention*, **38**, 472-481.

Bryant, J., and Zillmann, D. (2002). *Media Effects. Advances in Theory and Research. Second Edition.* Mahwah, NJ: Lawrence Erlbaum Associates.

Buss A.H., and Perry, M. (1992). The Aggression Questionnaire. *Journal of Personality and Social Psychology,* **63**, 452-459.

Chliaoutakis, J.E., Demakakos, P., Tzamalouka, G., Bakou, V., Koumaki, M., and Darviri, C. (2002). Aggressive Behavior While Driving as Predictor of Self-Reported Car Crashes. *Journal of Safety Research*, **33**(4), 431-443.

Cowan, J. Jr., Jones, B., and Ho, H. (2006). Safety Belt Use by Law enforcement Officers on Reality Television: A Missed Opportunity for Injury Prevention? *Journal of Trauma,* **61**, 1001-1004.

Dahlen, E.R., and White, R.P. (2006). The Big Five factors, sensation seeking, and driving anger in the prediction of unsafe driving. *Personality and Individual Differences*, **41**(5), 903-915.

De Pelsmacker, P., and Janssens, W. (2007). The effect of norms, attitudes and habits on speeding behavior: Scale development and model building and estimation. *Accident Analysis and Prevention*, **39**, 6-15.

Donovan, J., and Jessor, R. (1985). Structure of Problem Behavior in adolescence and young adulthood. *Journal of Consulting and Clinical Psychology,* **53**(6), 890-904.

DuRant, R.H., Rich, M., Emans, S.J., Rome, E.S., and Allerd, E. (1997a). Violence and weapon carrying in music videos. A content Analysis. *Archives of Pediatrics and Adolescent Medicine*, **151**(5), 443-448.

DuRant, R.H., Rome, E.S., Rich, M.R., Allred, E. Emans, J., and Woods, E.R. (1997b). Tobacco and Alcohol Use Behaviors Portrayed in Music Videos: A content Analysis. *American Journal of Public Health*, **87**(7), 1131-1135.

Edwards, E.D. (1991). The ecstacy of horrible expectations: Morbid curiosity, sensation seeking, and interest in horror movies. In B.A. Austin (Ed.), *Current research in film: Audiences, economics, and law* (Vol. 5, pp. 19-38). Norwood, NJ: Ablex.

Eggermont, S. (2006). *The impact of television viewing on adolescents' sexual socialization.* Leuven: School voor Massacommunicatieresearch. K.U.Leuven.

Fishbein, M. (2000). The role of theory in HIV prevention. *Aids care*, **12**(3), 273-278.

Fishbein, M., and Ajzen, I. (1975). *Belief, Attitude, Intention and Behavior. An introduction to Theory and Research.* Reading, Massachusetts: Addison-Wesley Publishing Company.

Fishbein, M., and Cappella, J.N. (2006). The Role of Theory in Developing Effective Health Communications. *Journal of Communication,* **56** (supplement), S1-S17.

Gerbner, G., and Gross, L. (1976). Living with television: The violence Profile. *Journal of Communication*, spring **26**(2), 172-194.

Gerbner, G., Gross, L., Morgan, M., and Signorielli, N. (1986). Living with Television: The Dynamics of the Cultivation Process. (pp. 17-40) In Bryant, J. and Zillmann, D. (Eds.) *Perspectives on Media Effects*. London: Lawrence Erlbaum Associates.

Glik, D., Kinsler, J., Todd, W., Clarke, L., Fazio, K., Miyashiro, R., Perez, M., Vielmetter, E., and Flores, R.C. (2005). Unintentional injury depictions in popular children's television programs. *Injury Prevention,* 11(4), 237-241.

Greenberg, B.S., and Akin, C.K. (1983). The Portrayal of Driving on Television. *Journal of Communication*, **33**(2), 44-54.

Greenberg, B.S., and Gregg, J.L. (1998). *The Portrayal of Safety Belts on Broadcast and Cable Television.* Departments of Communication and Telecommunication. Downloaded from website ACTS: Automotive Coalition for Traffic Safety Inc., Press Room http://www.actsinc.org/cableportrayal.cfm [accessed May 22, 2007].

Greenberg, B.S., and Thanki, D. (1997). *The Portrayal of Safety Belts in Hollywood Motion Pictures.* Departments of Communication and Telecommunication. Downloaded from website ACTS: Automotive Coalition for Traffic Safety Inc., Press Room http://www.actsinc.org/cableportrayal.cfm [accessed May 22, 2007].

Gruber, E.L., Thau, H.M., Hill, D.L., Fisher, D.A., and Grube, J.W. (2005). Alcohol, tobacco, and illicit substances in music videos: A content analysis of prevalence and genre. *Journal of Adolescent Health,* **37**(1), 81-83.

Gulliver, P., and Begg, D. (2004). Influences During Adolescence on Perceptions and Behaviour Related to Alcohol Use and Unsafe Driving as Young Adults. *Accident Analysis and Prevention*, **36**, 773-781.

Gulliver, P., and Begg, D. (2007). Personality factors as predictors of persistent risky driving behavior and crash involvement among young adults. *Injury Prevention*, **13**, 376-381.

Hall, A. (2005a). Audience Personality and the Selection of Media and Media Genres. *Media Psychology,* **7**, 377-398.

Hall, A. (2005b). Sensation Seeking and the Use and Selection of Media Materials. *Psychological Reports*, **97**, 236-244.

Hansen, C.H., and Hansen, R.D. (1990). Rock music videos and antisocial behavior. *Basic and Applied Social Psychology*, **11**(4), 357-369.

Hansen, C.H., and Hansen, R.D. (2000). Music and Music videos. In D. Zillmann and P. Vorderer (Eds.), *Media entertainment: The psychology of its appeal* (pp. 175-196). Mahwah, NJ: Lawrence Erlbaum Associates, Inc.

Harré, N. (2000). Risk Evaluation, Driving, and Adolescents: a Typology. *Developmental Review,* **20**(2), 206-226.

Haynes, C.A., Miles, J.N., Clements K. (2000). A confirmatory factor analysis of two models of sensation seeking. *Personality and Individual Differences*, **29**, 823-839.

Jacobson, H.A., Kreuter, M.W., Luke, D., and Caburnay, C. A. (2001). Seat Belt Use in Top Grossing Movies vs Actual US Rates 1978-1998. *American Journal of Public Health,* **91**(9), 1395-1396.

Jonah, B.A. (1997). Sensation seeking and risky driving: a review and synthesis of the literature. *Accident Analysis and Prevention*, **29**(5), 651-665.

Jonah, B.A. (1997). Sensation seeking and risky driving: a review and synthesis of the literature. *Accident Analysis and Prevention*, **29**, 651-665.

Krcmar, M., and Greene, K. (2000). Connections Between Violent Television Exposure and Adolescent Risk Taking. *Mediapsychology,* **2**, 195-217.

McNamara, L., and Ballard, M.E. (1999). Resting Arousal, Sensation Seeking, and Music Preference. *Genetic Social and General Psychology Monographs*, **125**(3), 229-250.

Pelletier, A.R., Quinlan, K.P., Sacks, J.J., Van Gilder, T.J., Gilchrist, J., and Ahluwalia, H.K. (2000). Injury prevention practices as depicted in G-rated and PG-rated movies. *Archives of Pediatrics and Adolescent Medicine*, **154**(3), 283-286.

Perloff, R.M. (2003). *The Dynamics of Persuasion. Communication and Attitudes in the 21st Century*. London: Lawrence Erlbaum Associates.

Perse, E.M. (1996). Sensation Seeking and the use of television for arousal. *Communication Reports*, **9**(1), 37-48.

Potts, R., Dedmon, A., and Halford, J. (1996). Sensation Seeking, Television Viewing Motives, and Home Television Viewing Patterns. *Personality and Individual Differences*, **21**(6), 1081-1084.

Reinhardt-Rutland, T.H. (2007). *Transportation safety issues portrayed on television: some thoughts on driver distraction.* Inj Prev [eLetter] 4 January 2007. http://injuryprevention.bmj.com/cgi/eletters/12/6/400#1626

Rich, M. Woods, E.R., Goodman, E., Emans, E., and DuRant, R.H. (1998). Aggressors or victims: Gender and race in music video violence. *Pediatrics,* **101**(4), 669-674.

Roskos-Ewoldson, D.R., Roskos-Ewoldson, B., and Carpentier, F.R. (2002). Media Priming: A synthesis. In Bryant, J. and Zillmann, D. (Eds.) *Media Effects. Advances in Theory and Research.* Mahwah, N.J.: Lawrence Erlbaum Associates.

Schierman, M.J., and Rowland, G.L. (1985). Sensation-seeking and selection of entertainment. *Personality and Individual Differences,* **6**(5), 599-603.

Sherman, B.L., and Dominick, B.L. (1986). Violence and Sex in Music Videos: TV and Rock 'n' Roll. *Journal of Communication,* Winter 1986, 79-93.

Smith, S.L., and Boyson, A.R. (2002). Violence in music videos: Examining the prevalence and context of physical aggression. *Journal of Communication*, **52**(1), 61-83.

Sommersflanagan, R., Sommersflanagan, J., and Davis, B. (1993). What's happening on music television – A gender role content analysis. *Sex Roles*, **28**(11-12), 745-753.

Sümer, N. (2003). Personality and behavioral predictors of traffic accidents: testing a contextual mediated model. *Accident Analysis and Prevention*, **35**, 949-964.

Sun, S.-W., and Lull, J. (1986). The Adolescent Audience for Music Videos and Why They Watch. *Journal of Communication*, winter 1986, 115-125.

Tapper, J., and Thorson, E. (1994). Variations in music videos as a function of their musical genre. *Journal of Broadcasting and electronic Media*, **38** (1), 103-114.

Ulleberg, P., and Rundmo, T. (2002). Risk-Taking Attitudes Among Young Drivers: the Psychometric Qualities and Dimensionality of an Instrument to Measure Young Drivers' Risk-Taking Attitudes. *Scandinavian Journal of Psychology*, **43**(3), 227-237.

Ulleberg, P., and Rundmo, T. (2003). Personality, Attitudes and Risk Perception as Predictors of Risky Driving Behavior Among Young Drivers. *Safety Science*, **41**(5), 427-443.

Van den Bulck, J., and Beullens, K. (2005). Television and music video exposure and adolescent alcohol use while going out. *Alcohol and Alcoholism*, **40**(3), pp. 249-253.

Van den Bulck, J., Beullens, K., and Mulder, J. (2006). Television and Music video exposure and adolescent 'alcopop' use. *International Journal of Adolescent Medicine and Health,* **18**(1), 107-114.

Weisskirch, R.S., and Murphy, L.C. (2004). Friends, porn, and punk: sensation seeking in personal relationships, Internet activities, and music preference among college students. *Adolescence*, **39**(154), 189-201.

Will, K.E., Porter, B.E., Geller, E.S., and DePasquale, J.P. (2005). Is Television a health and safety hazard? A cross-sectional analysis of at-risk behavior on primetime television. *Journal of Applied Social Psychology,* **35**(1), 198-222.

Wingood, G.M., DiClemente, R.J., Bernhardt, J.M., Harrington, K., Davies, S.L., Robillard, A., and Hook, E.W. (2003). A Prospective study of exposure to rap music videos and African American female adolescents' health. *American Journal of Public Health,* **93**(3), 437-439.

Winston, F.K., Woolf, K.D., Jordan, A., and Bhatia, E. (2000). Actions without consequences Injury-related messages in children's programs. *Archives of Pediatrics and Adolescent Medicine,* **154**(4), pp. 366-369.

World Health Organization. *World report on road traffic injury prevention – Main messages.* 2004. Downloaded from http://www.who.int/violence_injury_prevention/publications/road_traffic/world_report/factsheets/en/index.html (accessed 2 Oct 2007).

Zhang, Y., Miller, L.E., and Harrison, K. (2008). The Relationship between exposure to sexual music videos and young adults' sexual attitudes. *Journal of Broadcasting and Electronic Media*, **52**(3), 368-386.

Zuckerman, M., and Litle, P. (1986). Personality and curiosity about morbid and sexual events. *Personality and Individual Differences,* **7**(1), 49-56.

In: Music: Composition, Interpretation and Effects
Editor: Tamara E. Ivanova, pp. 75-90
ISBN: 978-1-60876-170-8

Chapter 4

EMPIRICALLY VERIFIED MUSIC BASED HEALTH INTERVENTIONS

Carol E. Smith[1,*], Marilyn Werkowitch[1], Ubolrat Piamjariyakul[1], Faye Clements[1], Helen Schaag[1] and Emily Dauz[2]
[1] University of Kansas School of Nursing, Kansas City, USA
[2] Menorah Medical Center, Kansas City, USA

Abstract

Music is known to have many benefits for health, ranging from inducing relaxation and sleep to improving a depressed mood. Music also enhances cognitive function such as attention to task, communication, learning, and memory. The objective of this chapter is to describe several empirically verified health strategies that are based on music. These strategies have been used across several clinical trials to assist individuals with managing serious chronic illnesses and to improve their health and safety.

Introduction

Music audiotapes can be used to teach and reinforce some important health and safety behaviors [1, 2]. Music with instructive and repeating lyrics can drastically improve memory and task attention, similar to the large effect of daily reminders [3, 4]. Music strategies are similar to focused imagery research, which has demonstrated a large positive effect size (.67) in adults' relaxation and improved cognition after 8 weeks of daily use. Further, research indicates that interactions improve between patients and their family caregivers when they listen to music [5]. It is widely touted that music binds groups of people together [6, 7].

Music has been shown to increase relaxation and can soothe patients as they begin new health interventions or treatments [8, 9, 10]. Listening to music with relaxing tempos and lyrics with encouraging instructive guidelines was found to be more pleasing to patients

[*] E-mail address: csmith@kumc.edu. Tel. (913) 588-1667, Fax (913) 588-1660. 3062 School of Nursing Building, 3901 Rainbow Boulevard, Kansas City, KS 66160-7502.

versus repetitious, verbal instructions [11]. Reduced anxiety and improved sleep also has been associated with music-based interventions [12, 13].

Conceptual Framework

The conceptual framework guiding the use of music for health strategies is based on the classic theory for reinforcing repetitive health behaviors or habits [14]. This behavioral theory, first defined by Triandis [15] and confirmed by others has consistently found that health behaviors can become habitual with positive reinforcement. Further, Triandis theory indicates that the repeated use of new health behaviors predicts the continued use of that behavior, which eventually results in long-term positive health outcomes [16, 17,18]. For example, if the new health behavior is a prescribed breathing exercise, which a music audiotape guides the person to do, then repeated use of the music audio-guide will lead to consistent daily use of breathing exercises and subsequently better breathing outcomes [19].

Music Strategies Tested in Clinical Trails

It is important to develop and then test these music audiotape guided strategies. Both CD-ROM and cassette tapes were developed to accommodate subjects' preference for use with a CD player or a cassette tape recorder. CD players were available for loan to subjects as needed for participation during the study. Working with various populations, our research team has used music in several of the strategies we have developed. In one study a music strategy used a music jingle to teach correct handwashing techniques. Another music guided strategy has resulted in a positive effect (d=.45) in adherence to prescribed sleep initiatives [20]. Lastly, music strategy was used for driver safety by reducing sleepiness of drivers [21, 22]. Development of these strategies and their testing results are described below.

Guiding Handwashing Techniques via Music

Handwashing is the first line of defense against disease producing microorganisms. The main purpose of washing hands is to cleanse the hands of bacteria or viruses and chemicals that can cause a person harm or disease. The Centers for Disease Control and Prevention (CDC) has indicated that it is well documented that the most important measure for preventing the spread of pathogens is effective handwashing. Furthermore, CDC has guidelines for handwashing techniques and how long to scrub vigorously to be able to cleanse the hands of microbes. Videotapes have shown that few people conduct hand-washing correctly or for length of time. Therefore, we developed, updated, and empirically verified the effectiveness of an audiotape handwashing music jingle titled, "All You Need for Good Handwashing: Water, Soap and Time" [23, 24]. The music in the audiotape is a 1-minute, easily hummed jingle. Repeating lyrics describe key handwashing techniques: cleaning under fingernails, removing jewelry, and vigorous scrubbing hands following the CDC recommended length of time of 20 to 60 seconds.

In the first study we tested the 1-minute audiotape handwashing jingle with 10 patients in order to establish that (1) patients would use the handwashing jingle and that (2) the jingle was easily memorized. We conducted this test with patients on Home Parental Nutrition (HPN) who have an intravenous catheter implanted, usually in the chest area, and are at high risk for developing an intravenous catheter infection. Intravenous catheter infection is the most frequent, life-threatening (33% mortality), and costly ($17,000 to $50,000) complication with an incidence up to 34%, whereas 2% is acceptable by infection control standards [25, 26, 27]. Prospective data indicate infections account for 13% to 75% of annual HPN complications, 60% of emergency room visits, and 42% of rehospitalizations [28, 29]. Major sources of intravenous catheter infections (40%) are skin and touch contamination with normal staph epidermis, bowel stoma e-coli, or strep fecalis, resulting in a mean of 1.2 to 4 sepsis yearly per HPN patient [30, 31]. HPN infection episodes are often due to known breaks in home aseptic technique or to poor handwashing [32 33]. Therefore, HPN patients who manage home care intravenous line infusions require strict handwashing for their safety [34, 35].

Patients were instructed to put soap on their hands and then start listening to the handwashing audiotape. Patients were then asked whether or not they rinsed their hands before the tape finished. Out of the 10 patients who responded, n=7 stated that they did rinse before the tape ended, indicating that they did not wash for a full minute. These patients were then asked to use the handwashing audiotape and to scrub their hands for the length of time the music jingle played. Patients were given a checklist log for one week and asked to record whether they sang the handwashing jingle or listened to the audiotape while washing their hands. One patient reported listening to the taped jingle and singing along while washing hands for the entire week. The majority of the patients (n=9) listened to the tape for one to two days and then reported they then sang the jingle from memory. This study established that patients would use the audiotape and that the handwashing jingle was easily memorized.

Videotaping Chronically Ill Patients Handwashing

In the next study we assessed patients by videotaping their handwashing. In this study 42 HPN patients agreed to be videotaped conducting their daily intravenous home care procedures including washing their hands. Prior to placing a video camera in patients' homes, families were assured that only the research team nurses would evaluate the videotapes. A release granting permission for use of the videotapes for educational purposes was signed by patients and family members prior to filming. CDC guidelines recommend handwashing with soap for 60 seconds to remove bacteria, depending on the amount of dirt and residue on the hands [36]. However, results of the videotaping of these 42 HPN patients before the had any instruction showed that only 13 (31.49%) met the CDC guidelines for scrubbing hands vigorously after having taken off all rings. Their range of handwashing time was from 14 to 137 seconds with the majority under 60 seconds. Therefore, two-thirds (69%) of the patients were not meeting CDC standards for length of handwashing per observation rating score.

In a subsequent follow-up study over a 12 week period, another n=38 HPN patients were instructed to use a checklist listing CDC handwashing guidelines and to chart their use of the music handwashing jingle. A total of 38 patients used this music for 2 weeks, 25 for 6 weeks, 20 for 10 weeks, and 18 for the full 12 weeks of the study, and the majority indicated they had memorized the music jingle. These patients had fewer subsequent infections ($p<.05$) after 6 months of using the handwashing jingle.

Singing about Handwashing in Multiple Languages for Community Health

The handwashing audiotape jingle was selected in a competitive blind review by the Kansas State Health Department for its $500,000 state-wide handwashing campaign for schools, public services, and restaurants [37, 38]. Timed videotaped data of the first use of this handwashing jingle used with 15 second graders revealed that these children washed their hands significantly longer (M=21 seconds; $p=.05$) while singing the song [39].

This handwashing jingle has also been translated into numerous languages and used by medical missions in six countries to teach handwashing techniques. The jingle and packet were translated (backward and forward techniques) into Spanish, and it was found to significantly ($p=<.05$) impove the length of handwashing by children (n=33) in remote villages in South America [40]. Hand hygiene can be accomplished in most remote settings using the handwashing jingle even without running water by using soap and pans of water (see Figure 1).

Hand hygienic behavior has been shown to cut the number of child deaths from diarrhea (the second leading cause of child deaths) by almost half and from pneumonia (the leading cause of child deaths) by one-quarter [41]. This handwashing jingle has been used and tested by others as well. An equipment company that provides all patients in Spain with HPN supplies requested permission to use these Spanish-version materials.

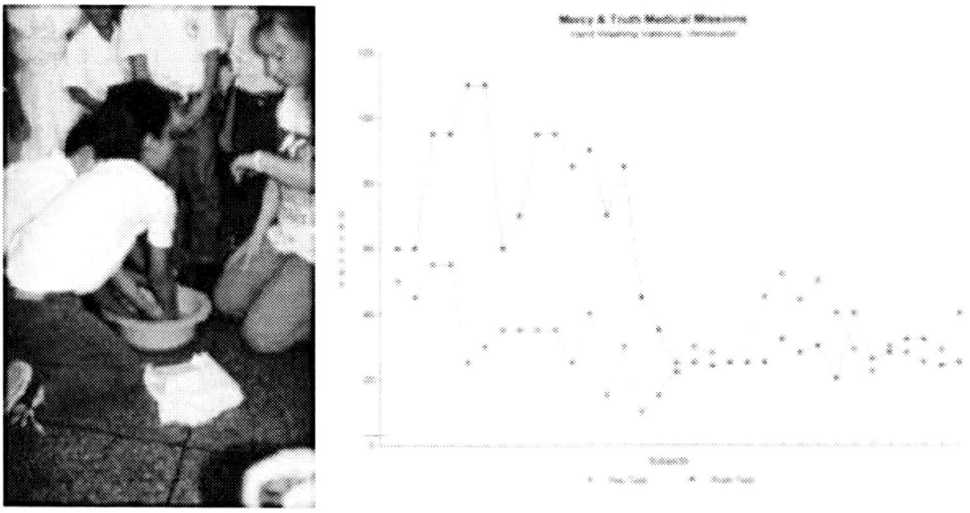

Figure 1. Timing handwashing of children while "Singing the Handwashing Song".

Summary Handwashing Jingle Testing Results

The handwashing jingle was found to be effective when used by school-aged children, in remote third world villages, and by patients with intravenous catheters. With the general public's recognition of the need to decrease transmission of microbes it continues to be important to provide good handwashing guidance [42, 43]. The studies with the music audiotape practice establishes easy-to-use repeated health behavior for effective handwashing in the public health domain. The handwashing jingle is being tested on the internet to cue handwashing activities and remind patients about daily scrubbing [44]. The music is easy to download, with even a basic modem, and the one-minute jingle is easily memorized.

Developing Music Strategies for Initiating Sleep

Getting a good night's sleep is easy for some people, but there are many more who suffer from sleep disorders, conditions that disturb sleep to the degree of interrupting daytime responsibilities [45]. It is estimated that over 70 million Americans suffer from one of the four most common sleep disorders: insomnia, sleep apnea, restless legs syndrome, and narcolepsy [46, 47].

In one study we focused a music strategy on patients with sleep apnea. This sleep disorder is characterized by pauses in breathing during sleep when the upper airway narrows and close as muscles relax naturally during sleep [48]. Each episode, called an apnea, lasts long enough so that one or more breaths are missed, and such episodes occur repeatedly throughout sleep. The apnea episodes reduce oxygen in the blood and cause arousal from sleep [49]. The most common treatment for sleep apnea is the use of a continuous positive airway pressure (CPAP) device, which "splints" the patient's airway open during sleep by means of a flow of pressurized air into the throat. Many patients are reluctant to use their CPAP machines because it requires wearing a nose or face mask or because of the clumsiness of the machine. National CPAP adherence rates are only 40 to 60% [50], yet with the repeated use of a CPAP machine patients will be more rested, alert, and less likely to have driving accidents related to sleepiness. Thus we developed an audiotape music strategy guiding restful sleep while using CPAP.

Music Strategy Guiding Nightly Sleep Initiation with CPAP

The feasibility of using a music audiotape with imagery as a nightly sleep strategy was tested in a clinical trial of 98 patients with sleep apnea who are required to use a CPAP machine each night. These patients were randomly assigned to an experimental or a control (no music strategy given) group. The sleep initiative with CPAP music intervention was used nightly for 12 weeks to instill a nightly sleep routine and a repeated procedure to place the breathing mask, tubing connections, and ventilation setting correctly. This music strategy was included in a kit with graphic photographs of peaceful sleep with CPAP and a audiotape titled "Imaging Your CPAP Restful Sleep" with soft-spoken directions for imaging sleep. [51]. This combination of step-by-step spoken audiotape instructions with music guides behavior so daily sleep routines become habitual [52].

The 10-minute audiotape guided the patient in imaging a restful sleep and created a relaxing environment by combining music and lyrics with a gradually decreasing music tempo. The slow tempo is synchronized with a resting heart rate cadence for enhancing relaxation [53]. The audiotape first instructs the patient to practice breathing in deeply and then exhaling slowly and deeply. Next, the patient is guided to relax his or her muscles slowly from head to toes, using repeated reminders for slow, deep breathing and spoken images of long restful sleep (e.g., "feel the gentle air pressure slowly filling your lungs") to sensitize patients to CPAP benefits [54] .

We tested patients' sleep while using this audiotape with an Acti-watch™, a non-invasive method of monitoring and recording gross motor movement while sleeping. Patients were instructed to wear it on the wrist of their non-dominant arm for the duration of their participation in the study. The data from the Acti-watch™ was later downloaded to a computer for analysis. Follow-up data revealed the experimental group used their CPAP significantly more than the control group [55]. All experimental patient ratings (100%) indicated that use of the audiotape helped them "get into the habit of using CPAP" every night. One patient suggested extending the "lovely-sounding soft-spoken voice" of instructions, adding other descriptive images of peaceful sleep, or varying the types of music on the audiotape.

Summary of Nighttime Sleep Initiation Music Strategy

Providing relaxing step-by-step audiotape instructions and soothing music can guide patients to manage the unfamiliar CPAP procedure, tasks, and equipment. We concluded that relaxing music engages patients in the repetitive behavior of using the CPAP machine, which is an important health habit. [56, 57]. Adherence to CPAP has shown to decrease use of other health services [58, 59]. This music strategy might also benefit persons with sleep disorders other than sleep apnea who find it difficult to fall asleep.

Music Strategies for Power Naps

The American Academy of Sleep Medicine recommends that adults get 7 to 8 hours of sleep a night, and the National Sleep Foundation (NSF) recommends 7 to 9 hours. As a society, we are a generally sleep-deprived nation due to the stress of daily living and the fast-paced lifestyle causing too little time to get the sleep we need. According to Michael J. Breus, PhD, long term clinical consequences of sleep deprivation include high blood pressure, heart attack and heart failure, stroke, obesity, depression and other mood disorders, and injury from accidents [60, 61]. Sleep deprivation effects on brain function include increased attention lapses, cognition deficits, memory loss, stress, and involuntarily falling asleep [62]. There is a high cost to society associated with sleep deprivation (daytime sleepiness) in lost productivity, occupational accidents, and loss of human lives [63].

According to the National Sleep Foundation, many people are alert in the morning with a natural dip in alertness in the mid-afternoon. This dip is due to a daily drop in body temperature. Short periods of sleep during the day restore cognitive abilities, and diligence to attend to detailed tasks (e.g., HPN, driving), and decrease daytime sleepiness (measured as alertness) [64]. "Daytime (during the afternoon) naps of less than 30 minutes, even those as brief as 10 minutes, restore wakefulness and promote performance and learning" [65]. Guided short naps have also been found to improve cognitive task performances such as attention to visual tasks [66].

Developing Music Strategies for Daytime Power Naps

We developed a music CD-ROM or audiotape guided imagery for taking a 10-minute power nap to help restore alertness and memory, relieve stress and daytime sleepiness, and reinforce the healthy behavior of taking a short day time nap. The power nap music strategy used an audiotape that guided a short, 10-minute nap using various types of music including classical, soft rock, spiritual, and country western [67]. The power nap audiotape promotes focused imagery of restful sleep that features a gradually slowing musical tempo with lyrics that guide deep breathing, muscle group relaxation, and sleep. As with the sleep initiation strategy, the gradually decreasing music tempo induces relaxation by synchronizing with a resting heart rate cadence [68, 69]. Studies directly comparing the effects of brief and longer daytime naps of 20 minutes, 30 minutes and 1 hour found that 10-minute naps decreased sleepiness and, improved alertness and reaction time measures significantly more than did the longer naps [70]. Subjects taking 10-minute naps continue to have significantly better measures of reaction time, alertness, cognitive performance, and vigilance after 2 to 3 hours than those sleeping longer than 10 minutes [71, 72]. Also, sleeping longer than 30 minutes can increase the risk of waking up feeling groggy and affect the ability to fall asleep at bedtime [73].

Short Naps for Fatigued Family Caregivers

Strategies for short 10-minute power naps, were taught to family caregivers members who assist ill loved ones at home. Sleep deprivation is a pervasive problem for family caregivers of loved ones on HPN due to nightly sleep interruptions from trouble-shooting infusion pump alarms and monitoring flow rates to prevent glucose and fluid overload. Even intermittent sleep deprivation is a risk to safety because of the effects on brain function, which include increased attention lapses, cognition deficits, memory loss, and involuntarily falling asleep [74, 75].

Early morning sleep disruption (which occurs during the daily 5 to 7 a.m. HPN infusion procedures) is associated with a decline in natural killer immune cells. Thus sleep deprived patients and caregivers have increased risks of infection [76, 77].

In this study HPN family caregivers (n=8) used a 10-minute power nap audiotape daily for two weeks. Caregivers were instructed to take a power nap using these guidelines:

1. Nap the same time every day, about 6 to 8 hours after waking up in the morning.
2. Preferably nap in your own bedroom, but can nap anywhere
3. Eliminate or minimize light by darkening or covering windows, or wear an eye mask.
4. Get comfortable, close your eyes, and start relaxing.
5. Turn on the music and follow the relaxation instruction for a peaceful sleep.
6. Do not try too hard to sleep, but concentrate on the feeling of relaxation. As you learn breathing techniques, you will move more quickly to sleep than when you first begin.
7. When the alarm goes off, get up. Don't continue to sleep.

Caregivers were reminded that naps are not a substitute for a full night of sleep. If they had a problem falling asleep or staying asleep at night after taking a power nap, then naps may not be for them. An Acti-watch^{TM}, as previously described was worn by the caregivers in order to measure their gross motor movements while they slept. This wristwatch uses actigraphy as a method of monitoring and recording gross motor movements. The movement graph data was later downloaded to a computer for analysis. Caregivers using the power nap music audiotape were shown on the wristwatch data to have relaxed movement, indicating they slept on average of 9.8 minutes during their music strategy nap. Caregivers reported that "these short daytime naps helped overcome exhaustion." These "recuperative daytime naps" provided caregivers with strategies to overcome the fatigue of taking care of an ill loved one which is important to their own health [78].

Short Naps for Graduate Students

Graduate students also often suffer from sleep deprivation due to their sleep pattern fluctuation to accommodate classes, work commitments, social activities, and cramming for tests. Short daytime naps decrease cognitive deficits, improve alertness, and consolidate learning of complex tasks [79]. A short daytime power nap can help alleviate irritability and exhaustion due to sleep deprivation in these students. Thus giving college students a method to remain alert during the day is essential to their learning and optimal performance [80]. We are beginning to study doctoral students use of the power nap CD-ROMs.

Recuperative Nap Strategies for Improving Safety on the Road

An NSF Sleep in America poll of 1000 workers found 32% admitted driving while drowsy at least once a month during the past year, and 2% experienced an accident or near accident because of driving while drowsy [81]. NSF recommends getting off the road if you are driving and start to feel drowsy. Driver fatigue and drowsiness result in reckless driving by failure to keep in the proper lane or running off the road. According to the Federal Motor Carrier Safety Administration (FMCSA), more than 750 people die and 20,000 more are injured each year due directly to fatigued commercial vehicle drivers [82]. For this reason, in April of 2003 the FMCSA instituted new hours of service regulations to curb truckers' problems with fatigue.[83]. While these laws have helped reduce the number of accidents, the U.S. National Transportation Safety Board still blames driver fatigue as a probable factor in 20 to 40% of truck accidents.

Music-guided recuperative naps have improved attention to monotonous tasks such as driving and to reaction time measured in milliseconds. Increased attention to driving tasks, improved decisions during driving as well as reduced distractibility and anxiety, have been associated with music-based strategies [84]. Short recuperative naps are strategies that can be used by drivers during on-duty breaks from driving to increase vigilance. Sleep laboratory studies of short recuperative naps have shown improvements in persons who had only 5 to 6 hours of sleep the night before the nap testing. Similarly, truck drivers report an average of 5.6 hours of on-duty sleep.

We undertook a study with truck drivers [85] to test the benefit of our music audiotape developed to guide truck drivers in short daytime naps. Also tested was whether drivers using recuperative naps and sleep safety strategies at least 3 times a week over a 3- month period would improve cognitive, fatigue, and health status measures.

The power nap strategy for truck drivers began in a single sleep laboratory setting. The truck drivers were instructed to stop at the sleep lab on their overland U.S. route. Participants in the study were 10 long-haul semi-trailer truck drivers including 2 women, with a mean age of 50.9 years. These truck drivers reported driving an average of 71,833 miles per year, with 2 of the 10 driving over 100,000 and 2 others over 150,000 miles annually. In this sample, 4 subjects had been drivers for less than 15 years, 3 between 16 and 20 years, and 3 for more than 30 years. None participated in team driving. The average time that subjects drove per week was 47.27 hours. Data collected using the AARP driving safety checklist data indicated that 60% of subjects reported near miss accidents "almost daily," and only 55% wore seatbelts. The majority of the near misses were reported as being caused by cars suddenly pulling in front of the truck. At baseline, only 2 of the subjects reported ever taking a nap or resting during their on-duty breaks, and only 1 of these 2 rested routinely. Subjects reported getting an average of 5.6 hours of sleep per night at home and 5 hours on the road (although reports varied from 2 to 8 hours). None of the subjects took medication for sleep routinely or the night before our sleep testing.

Prior to using the power nap, the truck drivers' vigilance was measured. Vigilance is defined as maintaining "attentiveness or alertness over time, during repetitive monotonous tasks." The Psychomotor Vigilance Task (PVT) measures attention to the task by having the driver press a button on the reaction time meter each time a flashing light randomly occurs during 6 minutes. PVT is the most common and reliable measure of vigilance used in the military and for commercial airline pilots. The sleep lab bedroom environment was prepared to be conducive for napping by using sound proofing and diminished lighting. Subjects were fitted with a LifeShirt™, a non-invasive monitoring system to track and record data on posture and movement as well as pulmonary and cardiac information. As with the Acti-watch™, the data was downloaded to a computer for analysis. After ensuring the drivers were comfortable with the LifeShirt™ on, the power nap music audiotape began. As previously described, the power nap audiotape consists of softly spoken instructions guiding subjects through breathing techniques focusing on relaxation toward a peaceful nap while music plays in the background. At the end of 10 minutes, an alarm with a repeated buzzing (though not jarring) sound woke the subjects.

Observation and recording data during the nap confirmed all participants were in a restful state during the nap. The majority of subjects (N=8) reported having slept during the nap. All subjects were readily aroused from the nap by the buzzing noise that is purposefully placed on the audio-recording to wake the driver after 10 minutes of music.

Vigilance Test Results

Pre- and Post-Nap vigilance results are shown in Figure 2. In the graph each **o** is one of the 72 to 78 times that reaction time was randomly tested. The PVT software calculates the slope of the line across the reaction times (measured in milliseconds). The reciprocal of the slope is calculated as a positive or a negative percent of the millisecond changes across the 6-

minute timeline. At pre-nap the drivers had negative percent change (M= -4.90, SD=3.56). This negative finding indicated that drivers' pre-nap reaction times continued to decline across the 6 minutes, thus their vigilance-to-task was poor. In contrast, at post-nap the percent change of reaction times across 6 minutes reverted to positive (M=4.35, SD=9.8). Thus, mean percent change (the measure of vigilance or attention to task) significantly improved post-nap (t=2.06, p=.032).

Pre- and post-nap PVT score comparisons were also tested with each driver matched as his own control. The mean 10% slowest and mean 10% fastest reaction times were calculated. The post-nap PVT recordings had significant mean improvements in the fastest mean reaction time (t= -2.45, p=.016) and decreases in the slowest mean reaction time (t=2.306; p=.021) over the pre-nap reaction time.

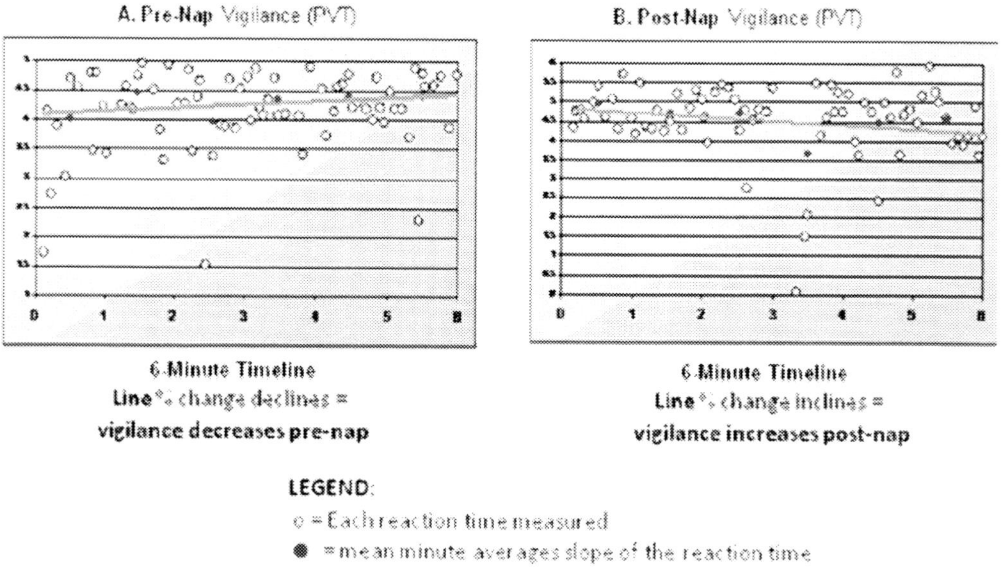

Figure 2. Driver vigilance as measured by Psychomotor Vigilance Task (PVT) pre-and- post nap test.

Confirmed Relationships between Vigilance and Driving Related Variables

The truck drivers self-rated tiredness was significantly decreased from pre- to post-nap on a scale from 1 to 10 (t=1.94, p=.003). There was a strong positive correlation (r=.683, p=.042) between post-nap percent change in vigilance (reaction time improvement across the 6 minutes) and the pre-nap self-rated tiredness. Pre-nap tiredness was negatively associated (r= -.761, p=.017) with post-nap mean fastest reaction time. In other words, the greater the pre-nap tiredness, the greater improvement in post-nap vigilance and reaction time. Investigators also found that there was an improvement in alertness and cognitive performance immediately following the 10-minute nap, that remained for up to 3 hours of post-nap testing [86, 87].

Three-Month Follow-up on Power Naps during On-duty Driving

Subjects were instructed to use the power nap audiotape over the next 3 months to take a daily 10-minute nap during on-duty rest periods. They also agreed to use a calendar to keep a record of daily on-duty naps. They were given instructions for setting up the cab or sleeping berth of the truck for comfort during the daily nap. All 10 truck drivers kept a calendar for 2 months and n=9 kept the calendar for all 3 months of the study. One driver's calendar indicated only 2 months of nap usage due to a 1 month layoff. Calendar data tabulation indicated that the 9 working truck drivers listed taking naps 81.3% of on-duty time. The drivers rated these short naps as restful 70.2% of the time across 3 months of calendar data. The woman driver used the naps 100% of the time. Overall, this music strategy of on-duty "recuperative" naps added to the few health promotion interventions available for truck drivers [88].

The cost to produce music strategy intervention packet (e.g., audiotape, handouts, placards, and diary) was between $10 and $40. Mailing the packets and telephone charges for instructions about using the music strategy averaged $5. A nurse was compensated $25 per patient to administer the intervention, which ranged in time from 45 to 60 minutes. The cost of the audiotape development was $1,000 ($7 charged across all patients for audiotape copy replication and packaging). Thus, the total cost of the experimental and placebo intervention administration per patient was $41 [89].

Conclusion

Music formed the basis for several health related strategies. Music can also engage patients by providing calm, clear instructions with pleasant, soothing lyrics to guide repetitive behavior and support daily habits to help patients continue their prescribed treatments. Each of the music audiotapes we developed and tested were used and positively evaluated by the study participants. The expected improvements from using these music strategies were achieved in each of these studies. Low cost and ease of use of these strategies supports continued use of music to improve health and safety [90]. Meta-analysis, clinical trials and other research continue to empirically verify music based health interventions [91, 92, 93, 94].

References

[1] Statum, M.J. (1988). Music for physical rehabilitation: An analysis of the literature from 1950-1986 and applications for rehabilitation settings. In C. E. Furman (Ed.). *Effectiveness of music therapy procedures: Documentation of research and clinical practice*. Washington, DC: National Association for Music Therapy, Inc., p. 81.

[2] Wolfe, D. (1988). Music therapy in group psychiatric work. In C.E. Furman (Ed.), *Effectiveness of music therapy procedures: Documentation of research and clinical practice*. Washington, DC: National Association for Music Therapy, Inc., p. 188.

[3] Furnham, A., Bradley, A. (1997). Music while you work: The differential distraction of background music on the cognitive test performance of introverts and extraverts. *Applied Cognitive Psychology*, **11**, 445-455.

[4] Stevens, T., Braithwaite, R.L., Taylor, S.E. (1998). Model for using hip-hop music for small group HIV/AIDS prevention counseling with African American adolescents and young adults. *Patient Education and Counseling*, **35**(2), 127-137.

[5] Balch, W.R., Bowman, K., Mohler, L.A. (1992). Music-dependent memory in immediate and delayed word recall. *Memory and Cognition*, **20**(1), 21-28.

[6] Why Music? 2008 *The Economist*, December 20; 41-43.

[7] White, J.M., (2000). State of the science of music interventions. *Critical Care Clinics of North America*, **12**, 219-225.

[8] Colt, H.G., Power, A., Shanks, T.G. (1999). Effects of music on state anxiety scores in patients undergoing firberoptic bronchoscopy. *Chest,* **116**, 819-824.

[9] Gerdner, L., Mentes, J., Titler, M. (1999). Individualized music intervention protocol. *Journal of Gerontological Nursing*, 10-16

[10] Raymond A.R., Macdonald, L. A., Mitchell, T., Dillon, M. G., Serpell, J. B., Davies, et.al. (2003). An Empirical Investigation of the Anxiolytic and Pain Reducing Effects of Music. *Psychology of Music*, April 31, 187-203.

[11] Koehler, J.A., Cha, J., Smith, C.E. (2002). Caregiving and aging: A model for using nursing interventions to meet the challenge of community health needs. *Home Health Care Management and Practice*, **14**, 127-135.

[12] White, J.M. (2001). Music as intervention: A notable endeavor to improve patient outcomes. *Nursing Clinics of North America*, **36**, 83-92.

[13] Lai, H. Good, M. (2005). Music improves sleep quality in older adults. *Journal of Advanced Nursing* **49**, 234-244.

[14] Godin, G. (1987). Importance of the emotional aspect of attitude to predict intention. *Psychology Reports*, **61**(3), 719-723.

[15] Triandis, H.C. (1997). *Interpersonal Behavior*. Monterey, CA: Brooks/Cole.

[16] Facione, N.C. (1993). The Triandis Model for the study of health and illness behavior: a social behavior theory with sensitivity to diversity. *Advanced Nursing Science*, **15**(3), 49-58.

[17] Smith, C.E. (2006). Support for adherence: A framework for Interventions. In J. Dunbar-Jacob, J. Erlen, E. Schlenk, and C. Stilley (Eds.) *Monograph: Methodological issues in the study of adherence*. Pittsburgh, PA: U of Pittsburgh Press, pp. 107-120.

[18] Lauver, D. (1992). A theory of care-seeking behavior. *Image - Journal of Nursing Scholarship*, **24**, 281-287.

[19] Smith, C.E., Dauz, E., Clements, F., Werkowitch, M., Whitman, R. (2009). Patient education combined in a music and habit-forming intervention for adherence to CPAP prescribed for sleep apnea. *Patient Education and Counseling*, **74**(2), p. 184-190.

[20] Smith, C.E., Dauz ,E., Clements, F., Cook, D., Doolittle, G., Leeds, W. (2006) Telemedicine improves adherence: Placebo trial, Telemedicine. *Telehealth Clinical Care and e-Health*, **12**(3), 289-297.

[21] Dinges, D.F. (1991). Abstract Preplanned cockpit rest: effects on vigilance performance in long haul flight crews. *Aviation Space and Environmental Medicine*, **62**(5), 451

[22] Smith, C., Clements, F., Schaag, H., Piamjariyakul, U., (2006) *Case finding in chronic illnesses.* KUMC Team research presentation.

[23] Noll, C. (1997). Student Reseach Grant from Sigma Theta Tau. Good Handwashing Technique "All You Need for Good Handwashing is: Water, Soap and Time". (Cassette Recording).

[24] Noll, C. (2004). Adhering to infection control procedures. University of Kansas. (Unpublished paper).

[25] World Health Organization. (2008). Global Handwashing Day: Planner's Guide. http://www.who.int/gpsc/events/2008/15_10_08/en/index.html

[26] Boyce, J.M. Boyce, J.M. Pittet, D. (2002). Guideline for Hand Hygiene in Health Care Settings Recommendations of the Healthcare Infection Control Practices Advisory Committee and the HICPAC/SHEA/APIC/IDSA Hand Hygiene Task Force. *MMWR Recomm Rep*, Oct 25, 51(RR-16), 1-45, quiz CE1-4

[27] Howard, L. (2006). Home parenteral nutrition: Survival, cost, and quality of life. *Gastroenterology*, **1**(30), S52-S59.

[28] Howard L, Ashley C. (2003) Management of complications in patients receiving home parenteral nutrition. *Gastroenterology*, **124**(6), 1651-1661.

[29] Delegge, M.H. (2002). Demographics of home parenteral nutrition. *Journal of Parenteral and Enteral Nutrition*, (5 Suppl), S60-S62.

[30] Eggiman, P., Pittet, D. (2002). Overview of catheter-related infections with special emphasis on prevention based on educational programs. *Clin. Microbiol. Infect.,* May 8(5), 295-309.

[31] Smith, C. E., Curtas, S., Kleinbeck, S. V. M., Werkowitch, M., Mosier, M., Seidner D. L., et al. (2003). Clinical trial of interactive and videotaped educational interventions to reduce catheter-related infection, reactive depression, and re-hospitalizations for sepsis in patients receiving home parenteral nutrition. *Journal of Parenteral and Enteral Nutrition*, **27**, 137-145.

[32] McGuckin, M., Waterman, R., Storr, I. J., Bowler, I. C., Ashby, M., Topley, K., et.al. (2001). Evaluation of a patient-empowering hand hygiene programme in the UK. *Journal of Hospital Infection,* **48**(3), 222-227.

[33] O'Grady, N. P., Alexander, M., Dellinger, E. P., Gerberding, J. L., Heard, S. O., Maki, D. G., et.al. (2002). Guidelines for the prevention of intravascular catheter related infections. Centers for Disease Control and Prevention. *MMWR* Recomm *Rep.* August 9, 51(RR-10), 1-29

[34] Center for Disease Control and Prevention. Accessed December 12, 2008 http://www.cdc.gov/Features/HandWashing//.

[35] Kleinpell, R. (2003). Advances in treating patients with severe sepsis. Role of Drotregocin Alfa (Activated). *Critical Care Nurse*, **23**(3), 16-29.

[36] Streed, S. A. (2003). Blueprint for success: The CDC guideline for hand hygiene in healthcare settings. *Extended Care Product* News *(ECPN)*, **86**(2), 14-20.

[37] Grady, P. (2003). Nursing research: making a difference. *Nursing Outlook*, **50**(6), 233-235.

[38] Werkowitch, M., Kleinbeck, S., Smith C. E. (2000). *Handwashing signs: A database intervention.* Presentation: Eight Kansas Nursing Research Exchange. University of Kansas, School of Nursing (CEU Program).

[39] Stephans, L. (1998). *The effect of a handwashing song and teaching module on the length of time spent handwashing in the preschool and school setting.* (Unpublished paper), University of Kansas

[40] Smith, C., (2004). Technology Home Care Research Institute Competitive Research Grant # 181027.

[41] World Health Organization. (2008). How to Handrub and How to Handwash Retrieved on 21 July, 2008. http://www.who.int/gpsc/events/2008/15_10_08/en/index.html

[42] World health Organization (2008). WHO Guidelines on Hand Hygiene in Health Care (Advanced Draft). Retrieved on 21 July, 2008. http://www.who.int/patientsafety/information_centre/Last_April_versionHH_Guidelines%5B3%5D.pdf

[43] Caburnay , C .A., Kreuter, M. W., Donlin, M. J. (2001). Disseminating effective health promotion programs from prevention research to community organizations. *Journal of Public Health Management and Practice,* **7**(2), 81-89.

[44] Smith, C.E. (2005). Technological Home Care: "Improving Home Caregiving". NIH Grant # NR-009078-01A1.

[45] Young, T., Peppard, P.E., Peppard, P.E., Gottlieb, D.J. (2002). Epidemiology of obstructive sleep apnea: A population health perspective. *American Journal of Respiratory Critical Care Medicine* , **165**, 1217-1239.

[46] Breus, J. 12. WebMD . Accessed December 12, 2008 http://www.webmd.com/sleep-disorders/guide/important-sleep-habits. Access 12/12/08.

[47] Giles T. L., Lasserson T. J., Smith B., White J., Wright J. J., Cates C. J. (2006). Continuous positive airways pressure for obstructive sleep apnea in adults. *Cochrane Database of Systematic Reviews*, Issue 3. Art. No.: CD001106. DOI: 10.1002/14651858.CD001106.pub3.

[48] Smith, C. E., Schaag, H., Wekowitch, M., Thompson, N., Puno, F. (2003). Family caregivers of patients using high-tech equipment. Presentation: Complex Family Technology. Johnson County Kansas. Community College (CEU Program).

[49] Kribbs, N.B., Pack, A.I., Kline, L.R. (1993). Effects of one night without nasal CPAP treatment on sleep and sleepiness in patients with obstructive sleep apnea. *American Review of Respiratory Diseases*, **147**, 162-1168.

[50] Silverberg, D. S., Iaina, A., Oksenberg, A. (2002). Treating obstructive sleep apnea improves essential hypertension and quality of life. *American Family Physicians,* **65**(2), 229-236.

[51] Noll, C. (1997). Student Reseach Grant from Sigma Theta Tau. Good Handwashing Technique "All You Need for Good Handwashing is: Water, Soap and Time". (Cassette Recording).

[52] Hsu, W., Lai, H., (2004). Effects of music on major depression in psychiatric inpatients. *Archives of Psychiatric Nursing*, **18**, 193-199.

[53] Lewis, K. E., Seale L., Bartle, I. E., Watkins, A. J, Ebden, P. (2004). Early predictors of CPAP use for the treatment of obstructive sleep apnea. *Sleep*, **27**, 134-138.

[54] Tusek, D., Church, J.M., Fazio, V.W. (1997). Guided imagery as a coping strategy for perioperative patients. *Association of Operating Room Nurses Journal*, **66**, 644

[55] Smith, C.E., Dauz ,E., Clements, F., Cook, D., Doolittle, G., Leeds, W. (2006). Telemedicine improves adherence: Placebo trial, Telemedicine. *Telehealth Clinical Care and e-Health*, **12**(3), 289-297.

[56] Houts, P.S. Nezu, A.M. Nezu, C.M., Bucher, J.A., Lipton, A (1994). American College of Physician's Home Care Guide for Cancer: How to Care for Family and Friends at Home. *Philadelphia: American Society of Internal Medicine*, ISBN: 0943126304, 206-260

[57] Sarin-Gulian, A., Heliker, B., Gawlinski, A. (2006). The effects of music twice daily on various outcomes in intensive care patients receiving mechanical ventilation: Improving umbilical venous catheter care by implementing an evidence-based practice guideline in the neonatal intensive care unit. *American Journal of Critical Care*, **15**, 84.

[58] Ronald, J., Delaive, K., Roos, L., Manfreda, J., Bahammam, A., Kryger, M.H., (1999). Health care utilization in the 10 years prior to diagnosis in obstructive sleep apnea syndrome patients. *Sleep*, **22**, 225-229.

[59] Weaver, T., Maislin, G., Schwab, R., Cantor, C., Younger, J., Dinges, D., et al. (2001). Perception of OSA risks and benefits and volition to use CPAP. *American Journal of Respiratory Critical Care Medicine*, **163**(Suppl. 5), A838.

[60] Quan, S.F., Gersh, B.J. (2004). Cardiovascular consequences of sleep-disordered breathing: Past, present and future: Report of a workshop from the National Center on Sleep Disorders Research and the National Heart, Lung, and Blood. *Circulation*, **109**, 951-957.

[61] Shah, N.A., Botros, N.A., Yaggi, H.K., Mohsenin, M.V. (2007). *Obstructive Sleep Apnea is Associated with an Increased Risk of Coronary Artery Disease and Death" (Session B12; Abstract # 1090). New Haven Ct: American Thoracic Society, A295*

[62] Sarkamo, T., Tervaniemi, M., Laitinen, S., Forsblom, A., Soinila, S., Mikkonen, M., et. al. (2008). "Music listening enhances cognitive recovery and mood after cerebral artery stroke". *Brain*, **131**(3), 866-876.

[63] Sassani, A., Findley, L..J., Kryger, M., Goldlust, E., George, C., Davidson, T.M. Reducing motor-vehicle collisions, costs and fatalities by treating obstructive sleep apnea syndrome. *Sleep*, **27**, 453-458.

[64] Takahashi, M., Arito, H. (2000). Maintenance of alertness and performance by a brief nap after lunch under prior sleep deficit. *Sleep*, **23**, 813-819.

[65] Alspach, G., (2008). Napping on the night shift: Slacker or savior? *Critical Care Nurse* http://ccn.aacnjournals.org

[66] Dinges, D.F. (1991). Abstract Preplanned cockpit rest: effects on vigilance performance in long haul flight crews. *Aviation Space and Environmental Medicine*, **62**(5), 451.

[67] Smith, C.E. (2005). Technological Home Care: "Improving Home Caregiving", NIH Grant # 1R01NR-009078-01A1 NR#009078A.

[68] Bonnet M.H. (1991). The effect of varying prophylactic naps on performance, alertness and mood throughout a 52-hour continuous operation. *Sleep*, **14**(4), 307-315

[69] Gerdner, L., Mentes, J., Titler, M. (1999). Individualized music intervention protocol. *Journal of Gerontological Nursing*, 10-16.

[70] Abrams, D.B., Orleans, C.T., Niaura, R.S., Prochaska, J.O. , Velicer, W., (1996). Integrating individual and public health perspectives for treatment of tobacco dependence under managed health care: A combined stepped-care and matching model. *Annals of Behavioral Medicine*, **18**(4), 290-304.

[71] Rowan, N. (2002). Short 'power naps' found best performance booster. *Reuters Health*, **14**(1), 35-40.

[72] Tietzel, A. J., Lack, L.C. (2001). The short-term benefits of brief and long naps following nocturnal sleep restriction. *Sleep*, **24**(3), 293-300.

[73] Tietzel, A.J., Lack, L.C. (2002). The recuperative value of brief and ultra-brief naps on alertness and cognitive performance. *Journal Sleep Research*, **11**(3), 213-218.

[74] Dinges, D.F. (1992). Probing the limits of functional capability: The effects of sleep loss on short-duration tasks. In R. Broughton and R. Ogilvie (Eds.), Sleep Arousal and Performance New York: Birkohauser-Boston, Inc., pp. 176-188.

[75] Jewett, M.E., Dijk, D.J., Kronauer, R.E., Dinges, D.F. (1999). Dose response relationship between sleep duration and human psychomotor vigilance and subject alertness.. *Sleep*, **22**(2), 171-178.

[76] Dinges, D.F., Douglas, S.D., Hamarm, S., Zaugg, L., Kapoor, S. (1995). Sleep deprivation and human immune function. *Advances in Neuroimmunology*, **5**, 97-110.
[77] Irwin, M., Mascovich, A., Gillen, J.C., Willoughby, R., Pike J., Smith, T. (1994). Partial sleep deprivation reduces natural killer cell activity in humans. *Psychosomatic Medicine*, **56**, 493-498.
[78] Smith, C.E. (2005). Technological Home Care: "Improving Home Caregiving", NIH Grant # 1R01NR-009078-01A1 NR#009078A.
[79] Mednick, S.C., Nakayama, K., Cantero, J.L., Atienza, M., Levin, A.A., Pathak, N. et.al. (2002). The restorative effect of naps on perceptual deterioration. *Nature Neuroscience,* **5**(7), 677-681.
[80] Smith, D.S. (1991). A comparison of group performance and song familiarity on recall tasks with older adults. *Journal of Music Therapy,* **28**(1), 2-13.
[81] Alspach, G., (2008). Napping on the night shift: Slacker or savior? *Critical Care Nurse* http://ccn.aacnjournals.org
[82] Advocates for Highway and Auto Safety (2008). Truck driver fatigue, accessed December 22, 2008. http://www.saferoads.org/issues/fs-truckdriverfatigue.htm
[83] Federal Motor Carrier Safety Administration (2005). Cut large truck fatalities in half by 2010. http://www.sharetheroadsafely.org.
[84] Morton, L. L., Kershner, J. R., Sieger, L. S. (1990). The potential for therapeutic applications of music on problems related to memory and attention. *Journal of Music Therapy*, **27**(4), 195-208.
[85] Smith, C., Clements, F., Schaag, H., Piamjariyakul, U., (2006). KUMC Team research presentation. Case finding in chronic illnesses.
[86] Bonnet, M.H, Arand, D.L. (1994). Impact of naps and caffeine on extended nocturnal performance. *Physiology and Behavior*, **56**, 103-109.
[87] Jewett, M.E., Dijk, D.J., Kronauer, R.E., Dinges, D.F. (1999). Dose response relationship between sleep duration and human psychomotor vigilance and subject alertness. *Sleep*, **22**(2), 171-178.
[88] Taylor, A.H., Dorn, L. (2006). Stress, fatigue, health, and risk of road traffic accidents among professional drivers: the contribution of physical inactivity. *Annual Review of Public Health*, **27**, 371-391.
[89] Moore, J., Smith, C.E., Spaniol, R. (2004). The dollars and sense of CPAP internet treatment. *Advances for Managers of Respiratory Care*, **13**, 50-54.
[90] 90 American Academy of Sleep Medicine. (2000). Cost justification for diagnosis and treatment of obstructive sleep apnea. Position statement of the American Academy of Sleep Medicine. *Sleep,* **23**, 1017-1018.
[91] Cepeda, MS., Carr, DB., Lau, J., alverez, H. (2006). Music for pain relief. *Cochrane Database Syst Rev*. CD004843.
[92] Leardi, S., Pietroletti, R., Angeloni, G., Neozione, S., Ranalletta, G., Del Gusto, B. (2007). Rondomized clinical trial examining the effect of music therapy in stress response. *Br J Surg*. **94**, 943-947.
[93] Nilsson, U. (2008). The anxiety and pain-reducing effects of music interventions: a systmetaic review. *AORN J.*, **87**, 780-807.
[94] Werner, PD., Swope, AJ., Heide, FJ. (2006). The Music Experience Questionnaire: development and correlates. *J Psychol*, **140**(4), 329-345.

In: Music: Composition, Interpretation and Effects
Editor: Tamara E. Ivanova, pp. 91-103
ISBN: 978-1-60876-170-8

Chapter 5

BURNOUT AND MUSIC USE IN A HIGH-COGNITIVE DEMAND OCCUPATION

Teresa Lesiuk*
Frost School of Music, University of Miami,
Coral Gables, Florida

Abstract

There is stress in the computer information systems industry that affects work performance and employee wellness [1, 2, 3]. The nature of the business of creating information systems is competitive and frequently involves aggressive deadlines. The designers and developers of these systems are required to be highly innovative and efficient in the face of constant time pressures. The leading consequence of the stressors is reported as increased frustration which "negatively impacts a job that requires high levels of discipline, focus, and systematic thinking" [1, p. 73].

In contrast, evidence from psychological literature suggests that mild positive moods contribute to creative problem solving in task performance [4]. Increases in positive mood helped explain the improved work performance of computer information systems developers (CISD) when working with preferred music [5]. In a recent study examining the prevalence of personality type and music use [6] in two different CISD groups, there were twice as many individuals with introversion than with extraversion; a trend opposite to the general population of nearly a two to one ratio of extraverts to introverts. Those with introversion also had significantly greater negative mood than those with extraversion, with possible interference to optimal work performance. Introverts, more so than extraverts, are reported to make use of music for emotional regulation [7]. Thus, there is a need to account for the way in which music influences these individual differences and consequential work performance.

Greater evidence of the beneficial use of music in this high-stress occupation is needed and should be addressed by a) obtaining an objective productivity score, as compared to previously self-reported productivity measures, and b) accounting for differential music effects on productivity for extraversion and introversion, and c) accounting for the effect of a timed music-condition for work performance.

Research from organization science is currently emphasizing the role of daily moods in job behaviors, job satisfaction, and work outcomes [8], as well as the importance of enhancing self-adaptation, referred to as "thriving", in the workplace. [9]. Additionally, differences in

* E-mail: tlesiuk@miami.edu. Tel: 305-284-3650. 5807 Ponce de Leon, 201-D, Coral Gables, FL 33146,

personality type will determine differential mood and arousal needs and consequential work outcomes [10]. This article contributes to the research on affect in the organizational literature, and, is pioneering in music-organizational literature for music psychology and music therapy. Lastly, the findings have practical implications for management of high-cognitive demand, high-stress occupations.

The purpose of this chapter is to provide a rationale for music-use to alleviate adversary psychological stress and improve quality-of-work in a high-cognitive demand occupation. Presently, organizational psychology has deemphasized what in the past has been understood as job satisfaction and has placed its lens on the role of daily moods and their effects on work behaviors and performance. Adverse moods, as a result of burnout and stress negatively impact work performance of high-cognitive demand tasks. Music listening experiences are reported to induce positive moods and reduce the adverse effects of negative moods. Findings in regards to music-mood induction have been shown through both behavioral and neurological evidence (Blood and Zatorre, 2001). This chapter presents background literature of music mood-induction and, in turn, mood and work performance outcomes. The findings from a study of a high-cognitive demand occupation are discussed in light of the music-mood and mood-task performance literature.

Burnout in a High-cognitive Demand Occupation

Productivity, or "quality-of-work," in computer information systems design may be measured by the degree of innovation and usefulness of the software design. The ability to produce work that is both novel and appropriate is adversely affected by stress and negative moods. A more specific stress, referred to as *burnout*, has been found in software development teams to be related to high-cognitive requirements and high-learning requirements (Raghavan, Sakaguchi, and Mahaney, 2008; Sethi, Barrier, and King, 1999; Sonnentag, Brodbeck, Heinbokel, and Stolte, 1994). Burnout is defined as a prolonged response to chronic emotional and interpersonal stressors on the job, and is defined by three dimensions: exhaustion, cynicism, and professional inefficacy (Maslach, Jackson, and Leiter, 1996). In December 2002, the Cutter Information Technology (IT) Journal dedicated an entire issue to "Preventing IT Burnout." Guest editor, Ed Yourdon stated:

> "Burnout is still a topic that most senior managers would rather not confront, but it has become so prevalent and severe that some IT organizations have become almost completely dysfunctional." (p. 2)

Mild positive feelings have been found to influence beneficially both creative and systematic problem-solving and task performance. A beneficial effect of music on task performance may be explained by increases in state positive affect (i.e. mild positive feeling) (Isen, 2005). A dopaminergic theory helps explain the neurophysiological mechanism of the affect enhancement (Ashby, Isen, and Turken, 1999). Increased dopamine activity in the anterior cingulate improves several cognitive abilities including long-term episodic memory, working memory and neural areas of problem solving and cognitive flexibility. When music evokes a pleasant mood and an increased arousal state, participants perform better on non-musical tasks. Thus, there is support for an increase in creative problem solving and task performance by workers, in this case, computer-based systems developers.

Computer information systems developers create application software to deliver business, science or educational information to users in an organization. The designers and developers of these systems are required to be highly innovative and efficient in the context of aggressive deadlines (personal communication with project managers, 2001-2009). Information technology work stressors have been identified, as well as the consequential stress on the employees, and the consequential detriments to their work (Longenecker, Schaffer, and Scazzero, 1999; Wastell, and Newman 1993).

> ...Design, the researchers say, is particularly vulnerable to stress-caused errors. They found that 42% of all design faults were directly attributable to programmer stress. Why? ... deep thinking, such as searching for solutions in a huge problem space... is required in the design phase... Deep thinking is easily affected by stress, which causes imperfect investigation. (Fujigaki, 1993, p. 34)

Kaluzniacky (2004) administered a *Stress Assessment Questionnaire* to 191 information systems developers in Canada. The questionnaire measured work-environment factors such as full or part-time work, type of organization, rate of change in the technology and methodology. Personal factors included level of education, personality type (i.e, Myers-Briggs Psychological Type) and degree of emotional dependency on the job. Respondents also assessed their level of job stress, degree of severity of common stress symptoms and the degree of perceived contribution to the stress of commonly identified factors. Results of the survey indicated that 53% of IS workers considered their work stressful, 23% felt they were close to burnout, 51% found deadlines were not reasonable, 57% found rapid change was characteristic of their job and 63% indicated a significant to dramatic increase in stress level in the last eight years. Respondents suggested that rapid changes, the need to perform more with less, and unrealistic expectations of users were reasons why there was an increase in stress over the last few years. The survey also included workers reporting commonly experienced feelings of frustration, pride in accomplishments, being overwhelmed and experiencing anxiety. Further, this author suggested that personality-related strategies and "emotional intelligence" be gained by the developers in order to combat what is a commonly felt experience of daily IT stress (Kaluzniacky, 2002).

The leading consequence of stress among 187 information technology professionals working in over 20 large U.S. organizations was increased frustration. Longenecker et al. (1999) surveyed the IT professionals and concluded that the high degree of frustration negatively impacts a job that requires high levels of discipline, focus, and systematic thinking.

Significant levels of cathecholamine excretion were found in a study of physiological responses concerning performance and workload of software engineers (Rissler and Jacobsson, 1987). The researchers concluded that the price paid for the engineers' high workload and performance was the mobilization of high levels of adrenaline. Lastly, work design decisions are impeded by frequent occurrences of narrowed attention from stress. Wastell and Newman (1993) posit that adversary stress produces a vicious cycle of uneconomical and maladaptive cognitive processing.

The Beneficial Effects of Music Listening on Mood and Quality-of-work

Knight and Rickard (2001) explored the effect of sedative music on participants' subjective and physiological stress levels. Eighty-seven undergraduate students, separated into a control and an experimental group, were exposed to a cognitive stressor in which they had 12 minutes to prepare an oral presentation on a topic related to their area of study. The control group prepared for the task in silence, while the experimental group listened to the *Pachelbel Canon in D.* Heart rate, blood pressure, cortisol, salivary immunoglobin A levels (IgA) and self-reports of anxiety were measured at rest and following the presentation of the stressor. The experimental group also indicated their familiarity with the piece, liking of the piece and whether they found it relaxing. Results from the control group revealed that the stressor task caused significant increases in all measures except salivary IgA. The presence of the music in the experimental group suppressed significant increases in subjective anxiety, systolic blood pressure, and heart rate. The researchers concluded that music acts as an anxiolytic treatment, that is, an anxiety preventative or anxiety reducing measure.

This experiment is interesting on several levels. Despite the consideration and expectation that familiarity and music training would influence the effect of the sedative music on stress reduction, the actual findings indicated an opposite trend. Effects were more pronounced for those participants who were unfamiliar with the music and had no musical training. The researchers also found that those participants who were most highly stressed by the cognitive stress task benefitted most from the anxiolytic effect of the music. Lastly, modest physiological reactivity was prevented for those participants who did not perceive the task as stressful. The findings of this study are relevant when considering the aforementioned information technology stressors, whether they be perceived as mild or severely adverse.

State anxiety decreased when music was used prior to and throughout two computer programming tasks as performed by seventy-two students in an introductory computer programming course (Lesiuk, 2000). The students were assigned to a condition of (a) no music, or (b) music prior to, or (c) music prior to and during tasks requiring locating syntax and logic errors. Results indicated a statistically significant effect of the music listening on state anxiety levels. The group with music prior to and during the tasks experienced the least amount of anxiety. While there was no statistically significant difference between the music conditions for the programming scores, the group hearing the most music achieved the highest mean scores in both syntax and logic tasks.

Both short-term and long-term positive affect was increased with music listening via headset in an ecologically valid study with 56 professional computer information systems designers from four different IT environments located throughout North America (Lesiuk, 2005). The presence of music listening, with the condition that developers were allowed to listen when they wanted, improved their quality-of-work and the time spent on a task. The developers narrative responses revealed the value of music listening for positive mood change and for enhanced perception in design work. Similarly, Lesiuk (in press) reported beneficial effects of music listening for state positive and negative affect and self-report of quality-of-work with 24 computer information systems developers (CISDs) in a North American information technology company. The beneficial effects were attributed to improved state positive affect which has been reported to improve creative problem-solving ability (Isen, 2005), particularly influential in generative problem-solving tasks (Forgas, 1998; Forgas and

George, 2001, Kraiger, Billings, and Isen, 1989; Mano, 1992), such as those required in CISD.

Lesiuk, Pons and Polak (2009) examined personality type, trait mood and use of music listening by 32 professional CISDs from two different IT environments in south Florida. Prevalence of basic personality type preferences were measured with the Myers-Briggs Type Inventory (MBTI). Trait mood, also known as an individual's emotional disposition, was measured with the Multiple Affect Adjective Check List (MAACL). Results from the exploratory pilot survey indicated a significant prevalence of Introversion, Thinking, and Judging types in the work culture with greater negative trait mood for Introversion and Feeling types. Although the developers listened to music on average 2 hours per day, music listening trends by type were reported, including findings such as Extraverts listened to music twice as much than Introverts, and Feeling types twice as much as Thinking types. The researchers also reported a statistically significant correlation concerning their daily average music listening time and depression. This correlation suggested that the more negative affect the employees experienced, the more they listened to music to subdue their negative feelings. Sloboda and O'Neill (2001) report that in everyday uses of music, majority of listening episodes evoke a positive mood experience.

Purpose

The following study reports the work stressors, level of burnout, and the perceived influence of music-use on self and work performance in a high-cognitive demand occupation.

Method

Participants and Study Design

Twenty-four participants from a high-tech company in North America, completed a 3-week music listening study in which participant demographics, stress, burnout, state mood and quality-of-work, and narrative responses to a music questionnaire were recorded. This chapter reports data of stress, burnout, rating of music liking and narrative comments in regards to music use. Stress and burnout measures were administered prior to the 3-week study for the purpose of understanding the work stressors experienced by the computer information systems developers. A music questionnaire was administered at the end of the study. See Lesiuk (in press) for report of state mood and quality-of-work outcomes.

A music library of 65 CDs, set up in the vicinity of the developers, contained style categories of Alternative, Classical, Country, Heavy Metal, Light Jazz, New Age, Pop, and a category labeled Other (e.g. Folk, Spirituals). Developers could listen to the music within their office cubicle via a personal stereo or with headsets on their computer. They could listen to music of their own choosing either from their own collections or from the music library. Participants were instructed to "Listen to the music when you want to, as you want to" for weeks 1 and 3, with a minimum of 30 minutes of listening daily. There was no music listening in week 2.

Measures

Participant Demographic Questionnaire. A researcher-designed questionnaire was administered for the purpose of gathering information concerning age, gender, years of work experience, level of education, years of formal music education, and amount of current daily music listening at work. The demographic information was collected to help describe characteristics of the work culture.

Narrative Work Stress Questionnaire. Ivancevich's Stress Diagnostic Survey (1988), narrative section only, was administered prior to the three week study. Participants were asked to describe their most stressful experience at work in the past 12 months.

Maslach Burnout Questionnaire. The Maslach Burnout Inventory General-Survey, MBI-GS, (Maslach, Schaufeli, and Leiter, 2001) is a 16-item questionnaire measuring three subscale: Exhaustion, Cynicism, and Professional Efficacy. Exhaustion sub-scale assesses feelings of being emotionally overextended and exhausted by one's work, while Cynicism represents attitudes of disillusionment and indifference about work in order to gain distance from its exhausting demands. Cynicism is considered a dysfunctional attitude as it reduces the energy available for performing work and for developing creative work solutions. The Professional Efficacy sub-scale assesses feelings of competence and successful achievement in one's work.

Burnout is conceptualized as a continuous variable, ranging from low to high degree of experienced feeling. A high degree of burnout is reflected in high scores on Exhaustion and Cynicism and low scores on Professional Efficacy. The frequency that the respondent experiences the indicated feelings is related to each subscale assessed using a six-point, fully anchored response format. Scoring the MBI-GS involves computing the average rating on the 0-6 frequency ratings across the items within each of the three subscales.

Music Listening Questionnaire. Degree of music liking was measured on a researcher-designed Music Listening Questionnaire in which participants also rated (1=not at all to 5=very much so) the degree to which music listening relieved their work stress and the degree to which they felt music refreshed them for their next work task. A narrative section followed in which participants were asked how they thought the music may have influenced them at work.

Materials

A music library of sixty-five CD's was supplied by the researcher. Style categories of music in the library included Alternative, Classical, Country, Heavy Metal, Light Jazz, New Age, Pop, and a category labeled Other (e.g. Folk, Spirituals). Employees could also choose music from their own music libraries or playlists. Participants were also offered one of two types of headsets: ear buds (Jensen stereo ear buds, JB6) or a head set (JVC Stereo headphones, HA-33). Some employees chose to use their own headsets or ear buds.

Results

Participants

The 24 participants (male=15, female =9), ages 26 to 55 (*M*=39.12), were mainly computer information systems developers. Their work experience ranged from 1 to 25 years (*M*=7.99 years) and their formal musical education experience ranged from 0 to 10 (*M*=2.40, *SD* = 3.44). There was also a wide range of reported daily music listening time at work, from no time spent to nearly the entire day (*M*=55.76 min., *SD*= 84.80 min.).

Work Stress

Participants shared what they felt was the most stressful part of their jobs. Table 1 displays several narrative responses to the statement "Please describe your most stressful experience at work in the past 12 months" (Ivancevich and Matteson, 1988).

Themes in common expressed by the developers included time pressures, heavy work loads, and organizational changes affecting job security.

Table 1. Narrative comments of the computer information developers' work stress

Not having clear objectives.
Just having so much work that I have to work at.
Ramping up on 3 new technologies, while delivering the business.
We have had as many as 6 different programs on the go at the same time, each requiring in put from me. The stress comes from trying to manage each set of deadlines and have all code submitted on time.
Having to push a deadline and disappointing my management team due to an oversight by me.
The same stress everyday, every week, every month – deadlines – always more tasks.
Having to review/present deliverables which I have not had sufficient time to complete.
Tight schedule.
Having to see some colleagues leave.
Tried to beat the deadline – worked 10-11 hours a day.
Layoff activity.
Heavy case load and deadlines were compounded by the fact my wife was going through "chemo" treatments.
Organizational change.

Burnout

Table 2 shows the sub-scale burnout scores of the IT development group. The ratings are established by the measure from comparisons made to other North-American worker sub-groups. The three burnout sub-scales were measured with the Maslach Burnout Inventory - General Survey (1996). The sub-scales scores, particularly high in the emotional exhaustion and cynicism areas, reflect the work stress experienced and commented upon in this IT environment.

Table 2. IT Developers' Mean Scores and Degree of Burnout

MBI-GS Subscale	*M*	*SD*	Degree of Burnout
Emotional Exhaustion	3.24	1.52	High
Cynicism	2.56	1.31	High
Professional Efficacy	4.59	1.22	Moderate

Table 3. Music liking and music benefit rating

How much they liked the music	4.48
How much they felt music relieved stress	4.19
How much they felt music refreshed them for their next work task	4.14

Table 4. Narrative comments of the computer developers' influences of music use

Comments that indicated a beneficial use of music:

- Happier.
- I'm a total believer. I shall continue with it.
- Kept me in a better mood – more sociable.
- Helps me to focus, stay more alert.
- Made me feel happier.
- Calming, focus better, feel like I accomplished more – time seemed to move faster.
- Something to look forward to. Gave me a change during the day. Made the day go by faster and the particular task.
- Made work get done faster. Made the day go by faster. Seem to be in positive mood.
- Positively. Overall, it was useful in keeping me focused on my task list and away from listening in and being disrupted by my environment.
- Helped complete mundane tasks. Good distraction.
- Feel better.
- Music usually takes me into a different world. I can concentrate better at work.
- Inspiration and motivation.
- It seemed to make me more productive.
- Very positively, got more things done.
- Help to focus and block out noise.

Table 4. Continued

Comments that indicated an interference or neutral effect of music use:

- Minor influence. Sometimes it distracted my work.
- Short-term relief of stress, but the effect passed quickly.
- Certain type of music made me concentrate but with other types I had a hard time focusing.

Music Listening

A final music listening questionnaire, based on a 5-point Likert scale (1=not at all to 5=very much so) was administered at the end of the 3-week study. Table 3 shows the averaged scores for music liking and music benefit.

Although the ratings of music liking and benefit are high, there are limitations to be noted in what music listening can do when stress becomes overwhelming. For example, during a previous study (i.e., Lesiuk 2005), one participating company was temporarily laying off their developers. While that group of developers almost always listened to music during its working hours, the routine of doing so was interrupted by the layoffs. A developer reported she did not listen to music for two days during a music listening week in one study, due to the layoffs. Perhaps when stress is overwhelming developers may not reach for music as a calming agent or coping strategy, as though paralyzed temporarily by the stressors.

Table 4 displays the responses given to the question *"How do you think the music listening influenced you at work?"* Of the 19 out of 24 responses given, 84% indicated a beneficial use of music while working, while 16% indicated a neutral or slightly negative effect.

Discussion

Work Stress and Burnout

Themes that recur in the participants' narrative responses as to the most stressful part of their job include time pressures, unrealistic deadlines, volume of work, job insecurity from organizational changes, and home stressors coupled with work stress. Not only do these stressors create psychological and physiological problems for the developers during work hours, but there may be significant carry-over of the effects of work stress to home-life. Indeed, researchers (Leiter and Durup, 1996) have found significantly more spillover from work stress to home than from home stress to work. Of note, same or similar types of stressors are reported by Longenecker (1999). Further, as suggested by Wastell (1993) stressors of this kind have negative effects on cognitive tasks and decision making in the form of decreased attention and missed information. The developers' comments also revealed distress similar to that reported by Fujigaki (1993) who found stress taking on different forms throughout the life-cycle tasks. While the present study did not analyze the type of stress-mood responses to each systems life-cycle task, the narrative comments and interview descriptions appear to correspond with Fujigaki's findings.

Although stress and negative affect are not used synonymously in this chapter, current-day understanding of stress posits that stress is manifested in negative emotions (Lazarus, 1990). The presence of music listening appears to suppress negative moods as mentioned by one developer, *I found music helpful when I needed a break from what I was doing (i.e., work got too frustrating).* This comment reveals a potential distress resembling that reported by Fujigaki (1993) who found particular negative moods at different stages of systems development. The need to decrease negative moods is indisputable when 42% of all design faults are reported to be directly attributable to stress.

Conclusion and Future Research

A growing trend of improved mood and quality-of-work via music use has been established in the aforementioned line of studies (Lesiuk, 2000, 2005, in press; Lesiuk, Pons, and Polak, 2009). However, work performance measures to date have included an experimenter-designed computer task completed by students (Lesiuk, 2000) and industry specific tasks, differing in length of project and complexity, as required of professional computer information systems developers (Lesiuk, 2005). Both of these work performance measures have proven less than ideal measures of work productivity. The 2000 study included an experimenter-designed performance task evaluated by experts, but the population was a student group, and therefore the findings could not be generalized to human-performance factors of professional developers. Further, although studies were carried out with professional developers completing tasks, the quality-of-work measure was completed by self-report, and the tasks often varied in nature of task, length of time of task and complexity. Therefore a productivity measure that is evaluated by experts and completed by professional developers is needed.

A recent study examining the prevalence of personality type and music use of 32 computer developers revealed that the IT groups contained twice as many individuals with introversion than with extraversion (Lesiuk, Pons, and Polak, 2009). The ratio of introverts to extraverts presented a trend opposite to the general population; that of nearly a two to one ratio of extraverts to introverts. Those with introversion also had significantly greater negative mood than those with extraversion. Introverts, more so than extraverts, are reported to make use of music for emotional regulation (Chamorro-Premuzic and Furnham, 2007). Thus, there is a need to account for the way in which music influences these individual differences and consequential work performance (Lesiuk, 2008).

In summary, greater evidence of the beneficial use of music for mood alteration and work performance in this high-stress occupation is needed and should be addressed by a) obtaining an objective work performance score, and b) accounting for mood as a mediator between music use and work performance, and c) accounting for music-mood change as reported by introversion and extraversion.

Themes from the narrative responses indicating a beneficial effect of music included improved mood, improved mood leading to other improved work behavior, permitted a change from work routine, sped up work task time, improved concentration. Purposeful music listening, as defined by an awareness of present and desired mood state to achieve an optimal generative problem-solving state, should be examined. An awareness of one's current mood state requires what Kaluzniacky (2002) referred to as an emotional intelligence. An outcome of achieving such intelligence, for example, could be a purposeful use of music listening prior to a large project or used during break times may be beneficial for task performance and employee wellbeing. Lastly, empirical findings would help establish a theoretical model of music-use in high-cognitive demand occupations and further elucidate the role of music in organizational settings.

References

[1] Longenecker, C. O., Schaffer, C. J. and Scazzero, J. A. (1999). Causes and consequences of stress in the IT profession. *Information Systems Management,* **16**(3), 71-77.

[2] Sethi, V., Barrier, T., and King, R. C. (1999). An examination of the correlates of burnout in information systems professionals. *Information Resources Management Journal,* **12** (3), 5-13.

[3] Raghavan, V. V., Sakaguchi, T., Mahaney, R. C. (2008). An empirical investigation of stress factors in information technology professionals. *Information Resources Management Journal,* **21**(2), 38-62.

[4] Forgas, J. P., and George, J. M. (2001). Affective influences on judgments and behavior in organizations: An information processing perspective. *Organizational Behavior and Human Decision Processes,* **86**(1), 3-34.

[5] Lesiuk, T. (2005). The effect of music on work performance. *Psychology of Music,* **33**(2), 173-191.

[6] Lesiuk, T., and Pons, A., and Polak, P. (2009). Personality, mood, and music listening of computer information systems developers: Implications for quality-of-work. *Information Resources Management Journal,* **22**(2),83-97.

[7] Chamorro-Premuzic, T., and Furnham, A. (2007). Personaltiy and music: Can traits explain how people use music in everyday life? *British Journal of Psychology,* **98**, 175-185.

[8] Fisher, C. D. (2000). Mood and emotions while working: missing pieces of job satisfaction? *Journal of Organizational Behavior,* **21**, 185-202

[9] Spreitzer, G., Sutcliffe, K., Dutton, J., Sonenshein, S., and Grant, A. M. (2005). A socially embedded model of thriving at work. *Organization Science,* **16**(5), 537-549

[10] Rusting, C. L. (2001). Personality as a moderator of affective influences on cognition. In Joseph P. Forgas' *Handbook of Affect and Social Cognition* (pp. 371-391). Mahwah, N.J.: L. Erlbaun Associates.

Ashby, F. G., Isen, A. M., and Turken, A. U. (1999). A neuropsychological theory of positive affect and its influence on cognition. *Psychological Review,* **106**(3), 529-550.

Blood, A. J., and Zatorre, R. J. (2001). Intensely pleasurable responses to music correlate with activity in brain regions implicated in reward and emotion. *Proceedings of the National Academy of Science*, **98** (20), 11818-11823.

Chamorro-Premuzic, T., and Furnham, A. (2007). Personaltiy and music: Can traits explain how people use music in everyday life? *British Journal of Psychology,* **98**, 175-185.

Forgas, J. P. (1998). On feeling good and getting your way: Mood effects on negotiation, cognition and bargaining strategies. *Journal of Personality and Social Psychology*, **74**, 565-577.

Forgas, J. P., and George, J. M. (2001). Affective influences on judgments and behavior in organizations: An information processing perspective. *Organizational Behavior and Human Decision Processes,* **86**(1), 3-34.

Fujigaki, Y. (1993). Stress analysis: A new perspective on peopleware. *American Programmer,* **6** (7), 33-38.

Isen, A. M. (2005). A role for neuropsychology in understanding the facilitating influence of positive affect on social behavior and cognitive processes. In C. R. Snyder and Shane J. Lopez (Eds.) *Handbook of Positive Psychology*. Oxford University Press.

Ivancevich, J., and Matteson, M. T. (1988). *Stress diagnostic survey: Comments and psychometric properties of a multidimensional self-report inventory.* Houston, Texas: FD Associates.

Kaluziacky, E (2002). A personal growth approach to preventing IT burnout. *The Journal of Information Technology Management,* **15** (12), 39-45.

Kaluzniacky, E. (2004). *Managing psychological factors in information systems work: An orientation to emotional intelligence.* Information Science Publishing.

Knight, W. E., and Rickard, N. S. (2001). Relaxing music prevents stress-induced increases in subjective anxiety, systolic blood pressure, and heart rate in healthy males and females. *Journal of Music Therapy,* **38** (4), 254-272.

Kraiger, K., Billings, R. S., and Isen, A. M. (1989). The influence of positive affective states on task perceptions and satisfaction. *Organizational Behavior and Human Decision Processes,* **44**, 12-25.

Lazarus, R. S. (1990). Theory-based stress measurement. *Psychological Inquiry,* **1**(1), 3-13.

Leiter, M. P. and Durup, M. J. (1996). Work, home, and in-between: A longitudinal study of spillover. *The Journal of Applied Behavioral Science,* **32** (1), 29-47.

Lesiuk, T. (2000). The effect of music listening on a computer programming task. *Journal of Computer Information Systems,* **40**(3), 50-57.

Lesiuk, T. (2005). The effect of music on work performance. *Psychology of Music,***33** (2), 173-191.

Lesiuk, T. (2008). The effect of preferred music listening on stress levels of air traffic controllers. *Arts in Psychotherapy,* **35**, 1-10.

Lesiuk, T. (in press) The effect of music on mood and performance in a high-cognitive demand occupation. *Journal of Music Therapy*.

Lesiuk, T., Pons, A., and Polak, P. (2009). Personality, mood, and music use of computer information systems developers: Implications for quality-of-work. *Information Resources Management Journal,* **22** (2), 83-97.

Longenecker, C. O., Schaffer, C. J. and Scazzero, J. A. (1999). Causes and consequences of stress in the IT profession. *Information Systems Management,* **16**(3), 71-77.

Mano, H. (1992). Judgments under distress: Assessing the role of unpleasantness and arousal in judgment formation. *Organizational Behavior and Human Decision Processes,* **52**, 216-245.

Maslach, C., Jackson, S. E., and Leiter, M. P. (1996). *Maslach burnout inventory manual, 3rd edition*. Consulting Psychologists Press, Inc.

Maslach, C., Schaufeli, W. B., and Leiter, M. P. (2001). Job burnout. *Annual Review of Psychology,* **52**, 397-422.

Raghavan, V. V., Sakaguchi, T., Mahaney, R. C. (2008). An empirical investigation of stress factors in information technology professionals. *Information Resources Management Journal,* **4**(2), 38-62.

Rissler, A., and Jacobsson. (1987). Cognitive efficiency during high work load in final system testing a large computer system. In H. J. Bullinger and B. Shackel (eds.) *Human-Computer Interaction*, 355-360. Elsevier Science Publishers.

Sethi, V., Barrier, T., and King, R. C. (1999). An examination of the correlates of burnout in information systems professionals. *Information Resources Management Journal,* **12** (3), 5-13.

Sloboda, J. A., and O'Neill, S. A. (2001). Emotions in everyday listening to music. Partrik N. Juslin and John Sloboda (Eds.) *Music and emotion: Theory and research.* Oxford University Press.

Sonnentag, S., Brodbeck, F. C., Heinbokel, T. and Stolte, W. (1994). Stressor-burnout relationship in software development teams. *Journal of Occupational and Organizational Psychology,* **67**, 327-341.

Wastell, D., and Newman, M. (1993). The behavioral dynamics of information system development: *A stress perspective. Accounting, Management and Information Technologies,* **3**(2), 121-148.

Yourdon, E. (2002). Opening statement. *The Journal of Information Technology Management,* **15** (12), 2-4.

In: Music: Composition, Interpretation and Effects
Editor: Tamara E. Ivanova, pp. 105-116
ISBN: 978-1-60876-170-8

Chapter 6

Heavy Metal, Country Music, and Personality: Reconsidering the Stereotypes

Richard L. Zweigenhaft
Guilford College, Greensboro, NC

Abstract

Social scientists have used a variety of methods to study the personalities of fans of various music genres. Some psychologists and sociologists have performed in-depth studies that have included participant observation, interviews and the administration of paper and pencil surveys and personality profiles (for example, Arnett, 1996, studied fans of heavy metal music, and Adams and Sardiello, 2000, studied fans of the Grateful Dead). More recently, Rentfrow and Gosling (2003) developed the Short Test of Music Preferences (STOMP) which they have administered to large samples of undergraduates at the University of Texas, along with various assessments of the Big Five personality traits. They concluded that music preferences fall within one of four distinct clusters (reflective and complex music, intense and rebellious music, upbeat and conventional music, energetic and rhythmic music) and that those who give high ratings to each of the four have distinct personality patterns. Zweigenhaft (2008), using an updated version of the STOMP, and a more in-depth measure of the Big Five personality traits, found some support for the patterns in Rentfrow and Gosling's (2003) studies, but also found important differences in the personality profiles of those who liked certain of the music genres. He also found some clear differences between fans of different music genres that fell within the clusters. that Rentfrow and Gosling (2003) identified. The study described in this chapter seeks to look in greater depth at the personality patterns for fans of two kinds of music who are frequently characterized based on limited data, or are stereotyped based on anecdotal data: those who like heavy metal music and those who like country music. The findings indicate that although there are kernels of truth in the previous portrayals social scientists have made about the fans of these music genres, they are also inaccurate in some meaningful ways.

Introduction

Social scientists have studied music fans in various ways. Some have performed in-depth studies of fans of certain genres. Developmental psychologist Jeffrey Arnett, for example, attended heavy metal concerts, interviewed self-described “metalheads,” and then asked the metalheads and non-metalhead control groups to fill out various paper and pencil tests. He concluded that these young men and women (most of the metalheads in his sample were men) were extremely alienated, highly impulsive, high sensation seekers, and prone to rebellious and destructive acts (Arnett, 1996). Similarly, sociologist Rebecca Adams and her students studied fans of the Grateful Dead. They attended concerts, they used the classic sociological technique of participant observation, and they asked Grateful Dead fans to fill out surveys. She and her students concluded that those fans of the Grateful Dead who considered themselves Deadheads saw this as an important part of their identity, and they considered themselves part of a subculture that sought community; for many it was part of a spiritual quest (Adams and Sardiello, 2000).

More recently, Rentfrow and Gosling (2003) have developed a paper and pencil measure to assess music preferences, the Short Test of Music Preferences (STOMP). Using factor analysis on respondents’ ratings of 14 music genres, they concluded that music fans fall into one of four clusters: those who like reflective and complex music (blues, classical, folk and jazz), those who like intense and rebellious music (alternate, heavy metal and rock), those who like upbeat and conventional music (country, pop, religious and sound tracks), and those who like energetic and rhythmic music (electronic, rap/hip-hop, and soul/funk). After administering this test to two large samples of respondents (N=1704 and N=1383), and asking them to fill out the 44-item Big Five Inventory (this relatively brief test assesses Emotional Stability, Extraversion, Openness, Agreeableness, and Conscientiousness), they found clear differences in personality among those who preferred the four types of music that they had identified. For example, there was a significant positive correlation between ratings on the genres that make up the reflective and complex music cluster and Openness ($r=.44$ and $r=.41$), but a significant negative correlation between ratings of the upbeat and conventional music cluster and Openness ($r=-.18$ and $r=-.15$).

Rentfrow and Gosling (2007) also have found that people hold common stereotypes about the personalities of those who like these 14 music genres. They asked judges to rate the “prototypical music fan” of one of the 14 music categories by rating that hypothetical person on a number of measures, including a five-item personality scale (the Five-Item Personality Inventory, or FIPI). When they calculated interobserver agreement scores (using intraclass correlations) for the eight judges who rated the hypothetical target person who liked each music genre, they found an overall correlation of .21, with higher scores for heavy metal ($r=.31$) and for country ($r=.27$). That is, there was moderate agreement among the judges about the personality characteristics they attributed to prototypical heavy metal and country fans (Rentfrow and Gosling, 2007).

In an accompanying study, Rentfrow and Gosling (2007) compared these stereotypes with the actual personality scores of a sample of 87 undergraduates at the University of Texas. For this study, they used the Ten-Item Personality Inventory (TIPI), another short measure of the Big Five personality dimensions (two items assess each of the five dimensions). They found a moderate correlation between the stereotypes of heavy metal fans and the actual personality scores of heavy metal fans (r=.23), but a small and slightly negative correlation (-.04) between the stereotypes and the actual scores for country music. Their findings on "stereotype validity" lead them to conclude that "many of the music stereotypes contain a kernel of truth" (Rentfrow and Gosling, 2007, p. 16).

North and Hargreaves (1999) also found that people hold "normative beliefs" about the fans of different kinds of music. North has used a large online sample (N=36,000), in which he asked respondents to rate 104 different musical styles, and to take a brief personality assessment. He, too, found connections between music preferences and personality. Those who like "rock/heavy metal," he concluded, are likely to have low self-esteem, and to be creative but they are not likely to be hard-working, outgoing, or at ease. In contrast, those who like "country and western" music, he found, are likely to be hardworking and outgoing ("Music Tastes….." 2008).

In their 2003 study, Rentfrow and Gosling concluded that "future research that includes narrower facets of personality is needed to provide a finer grained picture of the effects of personality on music preferences"(Rentfrow and Gosling, 2003, p. 1251). In an attempt to provide such a "finer grained picture," Zweigenhaft (2008), administered the 240-item NEO-PI to a sample of 83 undergraduates at Guilford College (in Greensboro, North Carolina), and looked separately at the music genres within each of Rentfrow and Gosling's (2003) four dimensions. Whereas Rentfrow and Gosling (2003, 2006, 2007) and North and Hargreaves (1999, 2008) have used short versions of tests that assess the Big Five personality traits (a five-item test, a ten-item test, and a 44-item version), the much longer 240-item NEO-PI assesses not only the Big Five traits (Neuroticism, Extraversion, Openness, Agreeableness and Conscientiousness) but the six facets that make up each of these five traits (thus, 30 facets). The NEO-PI is a more valid and reliable test of the Big Five than the shorter versions and, as such, it provides a more "fine grained" picture of personality. Zweigenhaft (2008) found meaningful differences among those who liked the various music genres within each of Rentfrow and Gosling's (2003) four clusters. He also found that some personality dimensions, like Openness, were much more likely to correlate with music preferences than others, like Agreeableness, and he found that having the assessments for the facets as well as the Big Five traits allowed for much more nuanced portrayals of the personalities of fans of various genres of music.

A number of studies have examined the extent to which fans of heavy metal music are troubled. In a study of 247 high school students from two schools in South Australia, Martin, Clarke and Pearce (1993) reported that for the females in their sample (n=109) "significant associations appear to exist" between preferences for "rock/heavy metal" and suicidal thoughts, self-harming acts, and depression. Acknowledging that the relationships were correlational, not causal, the authors concluded that "when females do prefer rock/heavy metal they appear to us within the limits of this preliminary study to be more disturbed as a group claiming more suicidal thoughts and acts, scoring as significantly more depressed and delinquent, and coming from more difficult families" (1993, p. 534). In a study of 121 high schools students in a small town in the American Midwest, Scheel and Westefeld (1999) found that fans of heavy metal were more likely to score low on a measure of "reasons for living," and they were more likely to have suicidal thoughts than fans of four other genres of music (pop/mainstream rock, rap, alternative and country). They concluded that "preference for heavy metal music may be a 'red flag' for increased suicidal vulnerability," though they acknowledged that the real source of the problem was likely to lie in "personal and family characteristics" rather than the music itself (1999, p. 253). In a study of 164 Canadian adolescents, Schwartz and Fouts (2003) compared those who liked "heavy metal and hard rock" with those who liked "pop music." They found that those who liked "heavy metal and hard rock" were more likely to be moody, pessimistic, overly sensitive, discontented and impulsive, and were more likely to have had problems in family relationships.

Perhaps because there is less concern that fans of country music are likely to harm themselves or others than fans of heavy metal, country music fans have not been studied as extensively as heavy metal fans. Moreover, the music encompassed by the term "country" has changed considerably over time. Still, the early stereotypes of country music fans alleged that they were unsophisticated, uneducated, and from rural areas – they were often called by the derogatory term of "hillbillies." In *Creating Country Music: Fabricating Authenticity*, sociologist Richard Peterson characterizes the prejudices against country music by noting that in its early days it was often considered to be "music for morons" (1997, p. 6).

This study, an extension of the research reported in Zweigenhaft (2008), is in part an attempt to explore some of these earlier claims about fans of heavy metal and country music. It is based on a larger sample than the 2008 study (the sample now includes 103 undergraduates at the same small liberal arts college) and it attempts to examine in greater depth the personalities of two highly stereotyped groups of music fans: those who like heavy metal and those who like country music. Given that the previous studies have depended on very brief assessments of the Big Five, assessments that have not allowed for the examination of the components of each of these personality traits (the facets), this study uses the 240-item NEO-PI to look more carefully at the extent to which there is stereotype validity for the fans of heavy metal and the fans of country music.

Methods

Between January 2001 and December 2008, students taking certain psychology courses at Guilford College, a small liberal arts college in Greensboro, North Carolina, were given the option of taking the NEO-PI. About one fourth of the students in these classes chose to do so. This entailed taking the test on their own (this generally took between 45 minutes and an hour), scoring it (this took about 15 minutes), and then going over the meaning of the scores with the instructor (this usually took about 45 minutes). Most students used their NEO-PI scores as part of self-analyses they wrote for an Introduction to Personality course. All agreed to allow the instructor to use the NEO-PI data for this research on music preferences.

As of the fall of 2004, 49 students had taken the NEO-PI; some were still students at the college, but others had graduated or transferred. Those who were still Guilford students, and those for whom addresses could be obtained, were contacted and asked to take the STOMP. Twenty-one did so. Between the spring of 2004 and the December 2009, another 82 students took both the NEO-PI and the STOMP (as of the spring of 2004, all students took a revised version of the STOMP which included an additional 7 musical genres: bluegrass, funk, gospel, international, oldies, opera, and punk; an even more recent version of the STOMP includes reggae and new age). The study of heavy metal and country music fans described in this chapter included NEO-PI scores and STOMP scores for 103 current or former students.

The students fell into three groups: 59 were traditional-aged undergraduates; 25 were part of the college's adult education program, which means they were Greensboro residents and older than 23; and 19 were part of an "early college" magnet program for gifted high school students from throughout the city that allowed them to enroll in college courses even though they were still in high school. The average age for the early college students was 16.7; for the traditional aged students, it was 19.5; and for the adult students, it was 37.2. As is true of the psychology majors at the college, most were female (76 of the 103) and most were white (85 of the 103).

Results

Differences among the Three Groups

First, the ratings that these three groups of students gave for heavy metal and country were compared. As can be seen in Table 1, there were no significant differences among the three groups in their ratings of heavy metal; notably, for the entire group this genre received the lowest overall rating of the 21 genres assessed (it was the lowest rated genre for two of the three groups, and second to lowest for the other). The ratings of country music, however, reveal a significant difference, and the within-group comparisons, using Tukey tests, indicate that the key difference was between the adult students, who were more likely to rate country music high, than the other two groups, who were more likely to rate it low. The difference between the early college and traditional students was not significant.

We also compared the male and female students in their ratings of these two genres. There was no significant difference in the way they rated country music, but the male students gave significantly higher ratings to heavy metal than did the female students (see Table 1).

Therefore, in the results that follow, we will look first at the entire sample, and then we will look separately at the responses of the younger and older students (because there were no differences between the early college and traditional-aged students, we combined those two groups), and separately at the responses of males and females.

Table 1. Ratings of heavy metal and country music, by type of student and sex

	Early College (n=19)	Traditional students (n=59)	Adult students (n=25)	F	p<	r=
Heavy metal	2.79	3.39	2.88	1.02	.36, ns	.01
Country	3.05[a]	3.81[b]	4.96[a,b]	6.26	.003	.33
	Men (n=27)	**Women (n=76)**		**t**	**p<**	**r=**
Heavy metal	3.78	2.93		1.96	.05	.19
Country	3.59	4.08		1.13	.26	.11

a=difference between early college and adult students significant, $p<.05$.
b=difference between traditional students and adult students significant, $p<.05$.

Table 2. Correlations between ratings of heavy metal and personality traits

	All (N=103)	Early College and Traditional (n=78)	CCE (n=25)	z-score contrasts	Women (n=76)	Men (n=27)	z-score contrasts
Neuroticism	-.14	**-.22 (p<.05)**	.02	z=1.00, p<.16	-.08	-.28	
Anxiety	-.09	-.20	.17	z=1.55, p<.06	-.04	-.24	
Angry hostility	-.11	**-.26 (p<.05)**	.23	**z=2.06, p<.02)**	-.09	-.17	
Depression	-15	-.15	-.16				
Self-consciousness	-.06	-.04	-.15		-.03	-.12	
Impulsiveness	-.08	-.13	.06		-.08	-.08	
Vulnerability	-.12	-.16	-.07		-.06	-.24	
Extraversion	-.10	-.04	-.27		-.12	.07	
Warmth	-.16	-.04	**-.40**	**z=1.91, p<.03**	-.10	-.13	
Gregariousness	-.12	-.07	-.28		-.10	-.08	
Assertiveness	-.11	-.12	-.08		-.16	.08	
Activity	-.08	-.05	-.12		-.23	.35	**z=2.55, p<.01**
Excitement-seeking	.06	.09	-.08		.09	-.01	
Positive Emotions	-.01	.05	-.19	z=0.99, p<.16	.01	.11	
Openness	.15	.13	.14		.18	.26	
Fantasy	.09	.07	.07		.05	.27	
Aesthetics	.07	.06	.05		.13	.15	
Feelings	.09	.01	.26		.03	**.47**	**z=2.04, p<.02**
Actions	.08	.16	-.13	z=1.20, p<.12	.22	-.10	
Ideas	.11	.07	.16		.10	.13	
Values	.16	.14	.15		.19	.22	
Agreeableness	-.08	-.07	-.18		.04	-.24	
Trust	-.08	-.08	-.14		.02	.22	
Straightforwardness	-.09	-.14	.05		0	-.14	
Altruism	.05	.09	-.05		-.01	.27	z=1.22, p<.10
Compliance	-.11	-.14	-.07		-.16	-.18	
Modesty	-.10	-.04	-.27		.10	**-.35**	**z=1.98, p<.02**
Tendermindedness	-.05	-.04	-.15		.06	-.18	
Conscientiousness	-.09	-.10	0		-.17	.12	z=1.24, p<.10
Competence	.01	.05	-.03		-.07	.25	z=1.38, p<.08

	All (N=103)	Early College and Traditional (n=78)	CCE (n=25)	z-score contrasts	Women (n=76)	Men (n=27)	z-score contrasts
Order	-.08	-.13	.13	z=1.08, p<.14	-.21	.22	**z=1.86, p<.03**
Dutifulness	-.13	-.12	-.14		-.16	-.02	
Achievement striving	-.13	-.14	-.09		-.15	-.06	
Self discipline	-.03	-.03	0		-.10	.11	
Deliberation	-.04	-.07	.07		-.10	.18	

bold=p<.05.

Heavy Metal

Neuroticism. Although fans of heavy metal are often portrayed as neurotic and, especially, teeming with angry hostility, this was not the case for the students in this study. As can be seen in Table 2, the correlation between ratings on heavy metal and Neuroticism was negative, not positive (r=-.14). That is, the more these students liked heavy metal, the lower their score on Neuroticism. This correlation was even stronger, and statistically significant, for the younger students (r=-.22, p<.05; for the older students, it was quite close to zero at r=.02).

The same pattern held for angry hostility, one of the facets of Neuroticism most typically attributed to fans of heavy metal: for all students the correlation was negative (r=-.11), and for the younger students it was stronger and statistically significant (r=-.26, p<.05). In fact, the correlations were negative for all six of the facets that make up Neuroticism (including the facet of impulsiveness, r=-.08, another attribute that is often attributed to fans of heavy metal). Notably, for five of the six Neuroticism facets these negative correlations were stronger for the younger respondents than the older respondents.

There was some evidence, however, among the older respondents, to support the stereotype of heavy metal fans as Neurotic. For these students, the correlations between ratings of heavy metal and two of the six facets – anxiety and angry hostility – were positive, not negative (r=.17 and r=.23), and z score contrasts between the two correlations (for the younger students and the older students) were in one case statistically significant (z=2.06, p<.02) and in the other quite close (z=1.55, p<.06); see Table 2.

The other Big Five traits. The ratings of heavy metal correlated negatively with Extraversion (r=-.10), positively with Openness (r=.15), negatively with Agreeableness (r=-.08) and negatively with Conscientiousness (r=-.09). None of these correlations was statistically significant, but they do suggest a personality pattern that tends to be introverted, not especially prone to behaving agreeably or conscientiously, yet open to experience. Again, a separate look at the older students suggests that for them preferences for heavy metal may be accompanied by interpersonal difficulties – the more they liked heavy metal, the lower they scored on warmth (r=-.40, p<.05), gregariousness (r=-.28, p<.18) and positive emotions (r=-.19, p<.35).

Table 3. Correlations between ratings of country music and personality traits

	All (N=103)	Early College and Traditional (n=78)	CCE (n=25)	z-score contrasts	Women (n=76)	Men (n=27)	z-score contrasts
Neuroticism	.07	.21	-.26	**z=2.04, p<.02**	-.06	**.38**	**z=1.90, p<.03**
Anxiety	.11	**.28**	-.27	**z=2.40, p<.01**	-.02	**.48**	**z=2.23, p<.01**
Angry hostility	.08	.12	-.16		-.01	.27	
Depression	.10	**.24**	-.26	**z=2.17, p<.01**	-.01	.36	**z=1.60, p<.05**
Self-consciousness	.02	.16	-.17	z=1.41, p<.08	-.08	.29	**z=1.56, p<.05**
Impulsiveness	-.03	-.01	-.23		-.03	-.07	
Vulnerability	-.03	.05	-.12		-.13	.19	z=1.33, p<.09
Extraversion	.02	.03	-.01		.08	-.19	
Warmth	-.09	-.09	-.07		-.09	-.20	
Gregariousness	-.04	-.08	.05		-.01	-.25	
Assertiveness	.12	.10	.08		.15	.02	
Activity	.14	.13	.01		.13	.10	
Excitement-seeking	-.07	-.04	-.05		-.01	-.23	
Positive Emotions	-.04	.02	-.08		.05	-.30	z=1.48, p<.06
Openness	-.18	-.07	-.20		-.13	-.35	
Fantasy	**-.22**	-.07	**-.49**	**z=2.58, p<.01**	-.19	-.34	
Aesthetics	-.07	-.02	-.01		-.03	-.26	
Feelings	-.03	.05	-.08		0	-.18	
Actions	.02	.02	.09		.04	-.09	
Ideas	-.16	-.06	-.28		-.15	-.19	
Values	**-.25**	**-.22**	-.08		-.16	**-.50**	**z=1.60, p<.05**
Agreeableness	-.04	-.01	.04		-.14	.10	
Trust	-.12	-.14	.04		-.09	-.26	
Straightforwardness	-.05	-.06	.15		-.18	.19	z=1.54, p<.06
Altruism	.09	.13	-.12		.02	.19	
Compliance	-.12	-.18	.22	**z=1.73, p<.04**	-.13	-.06	
Modesty	.07	.17	-.11		-.10	.29	**z=1.64, p<.05**
Tendermindedness	-.10	-.01	-.14		-.20	.05	

	All (N=103)	Early College and Traditional (n=78)	CCE (n=25)	z-score contrasts	Women (n=76)	Men (n=27)	z-score contrasts
Conscientiousness	.14	0	**.43**	**z=1.95, p<.03**	.12	.20	
Competence	.12	.02	.30		.15	.05	
Order	.05	-.06	.15		.02	.18	
Dutifulness	.18	.10	.34		.12	.28	
Achievement striving	**.20**	.13	.35		.16	.26	
Self discipline	.08	-.07	**.43**	**z=2.25, p<.01**	.07	.13	
Deliberation	.04	-.06	**.43**	**z=1.98, p<.02**	.06	0	

bold=p<.05.

As can also be seen in Table 2, the internal analyses using z-score contrasts to compare the patterns for the male and female respondents indicated a number of significant differences. The men who liked heavy metal, more than the women who liked heavy metal, were likely to score high on activity, openness to feelings, competence and order, but scored lower on modesty. Gender, then, like age, is a mediating factor when it comes to the personality patterns of heavy metal fans

Even acknowledging the differences based on age and gender, these findings generally challenge the portrayal of heavy metal fans as neurotic, angry, or impulsive, and they suggest that the more students liked heavy metal the more open they were to experience. At the same time, these data indicate that the more they liked heavy metal the more likely they were to be introverted, less agreeable and less conscientious.

Country

Neuroticism. As can be seen in Table 3, for all students, the correlation between ratings for country music and Neuroticism was nonsignificant, but positive (r=.08). Similarly the correlations with anxiety, angry hostility, depression and self-consciousness were positive but nonsignificant; the correlations with the other two Neuroticism facets, impulsiveness and vulnerability, were small, negative, and nonsignificant.

Separate analyses based on the age of the students, however, revealed strikingly different patterns: for the younger students, the correlations for Neuroticism, anxiety, angry hostility and depression were positive, but for the older students these correlations were negative, and the comparisons based on z-score contrasts between the two sets of correlations were either statistically significant or approached significance. That is, the more the younger students liked country music, the more neurotic they were, but the more the older students liked country music, the less neurotic they were. These patterns were especially strong (and statistically significant) for the facets of anxiety and depression: the more the younger students liked country music, the more anxious and depressed they were; the more the older students liked country music, the less anxious and depressed they were.

When the data were analyzed based on gender, the correlations for women were small and negative but for men they were larger, positive, and the contrasts were significant or approached significance. That is, there was no indication for women that preference for country music was related to Neuroticism, but for the men preferences for country music was related to Neuroticism and, correspondingly, to most of the six Neuroticism facets. Z-score comparisons revealed that these correlations for the men were significantly different than those for the women.

Not surprisingly, given the differences based on age and gender, for the 19 students who were both male and younger, preferences for country music correlated quite strongly with certain of the Neuroticism facets (e.g., for anxiety, $r=.48$, $p<.04$, and for depression, $r=.42$, $p<.07$).

The other Big Five traits. There were no meaningful patterns related to Extraversion, but those who liked country music were likely to score low on Openness ($r=-.18$). They were especially likely to score low on openness to fantasy ($r=-.22$, $p<.05$), openness to ideas ($r=-.16$) and openness to values ($r=-.25$, $p<.05$); notably, although the correlations on most of the six Openness facets for both the men and the women were negative, the correlations for the men were stronger, and on openness to values this difference was statistically significant ($r=-.16$ for the women, $r=-.50$ for the men; $z=1.60$, $p<.05$). There was a slight negative correlation between ratings of country music and Agreeableness; the only Agreeableness facet of interest was compliance, on which the younger respondents tended to score low and the older respondents tended to score high.

The higher the ratings for country music, the higher the score on Conscientiousness ($r=.14$), and this pattern was strongest on the facets of dutifulness and achievement striving. The internal analyses indicated that the older students who liked country music were especially Conscientious, dutiful, and achievement striving, and they were also self-disciplined and deliberate in their thinking and actions.

These findings suggest that preferences for country music correlated with Neuroticism – especially for the younger students, and for the male students. Those who preferred country music tended to be closed minded, but they were also conscientious and tended to score high on achievement striving.

Conclusions

These findings, based on personality assessments of much greater depth than many of the previous studies, call into question the stereotype validity of heavy metal and country music fans. Although, as previous researchers have claimed, there are kernels of truth in the stereotypes of heavy metal fans and country music fans, the fact that there are such strikingly different patterns based on both age and gender demonstrate the risk of buying into the stereotypes, or of overgeneralizing from data sets without controlling for age and gender (and, perhaps, other factors such as culture and geography).

In particular, these findings challenge previous portrayals of heavy metal fans as neurotic, angry, and impulsive. They do provide some support for the view that fans of country music are traditional in their views (they are not open to experience, not especially open to new ideas, nor are they open to questioning traditional values), and they indicate that they are both agreeable and conscientious.

The internal analyses demonstrate that on some personality dimensions both age and sex mediate in meaningful ways, and therefore broad generalizations based on music preferences should acknowledge that younger and older, and male and female, fans of heavy metal and country differ. This was especially the case for older and younger fans of country music in terms of Neuroticism, and especially the facets of anxiety and depression – for the younger respondents, the more they liked country music the more Neurotic, anxious and depressed they were, but for the older respondents, the more they liked country music the less Neurotic, anxious and depressed they were. It is also possible that when younger and older students are asked to rate heavy metal music, or country music, they are not thinking of the same music (the older respondent, for example, may be thinking of heavy metal music as it was in the early 1990s, but the younger respondent might be thinking of heavy metal ten or fifteen years later; similarly, the older respondents might think of country music as it was defined twenty or thirty years ago, and the younger respondent might think of country music as it is much more broadly defined today).

Finally, it is important to keep in mind the distinction between those who express preferences for music genres (as in this study), and those who consider a certain kind of music to be important enough to them to be part of their identity (as in the case of self-identified metalheads, or Deadheads). The patterns observed in Arnett's (1996) research on metalheads, or the work described in Adams and Sardiello (2000) on Deadheads, may be more pronounced because the level of identity with the music for those they studied was more substantial. To be a fan of the Grateful Dead does not necessarily mean one is a Deadhead, and to like heavy metal music does not mean one is a metalhead. Arnett (1996, p. 179) correctly distinguishes between a "taste culture" and a "subculture," and this study has examined the former.

References

Adams, R. G. and Sardiello, R. (Eds.) (2000). *Deadhead Social Science: "You Ain't Gonna Learn What You Don't Want to Know."* Walnut Creek, CA: Alta Mira Press.

Arnett, J. J. (1996). *Metalheads: Heavy Metal Music and Adolescent Alienation.* Boulder, CO: Westview Press.

Martin, G., Clarke, M. and Pearce, C. (1993). Adolescent suicide: Music preference as an indicator of vulnerability, *Journal of the American Academy of Child and Adolescent Psychiatry*, **32**, 530-535.

"Music Tastes Link to Personality," BBC News, December 13, 2008.

North, A. C. and Hargreaves, D. J. (1999). Music and adolescent identity, *Music Education Research*, **1**, 75-92.,

North, A. C. and Hargreaves, D. J. (2008), *The social and applied psychology of music.* New York: Oxford University Press.

Peterson, R. A. (1997). *Creating Country Music: Fabricating Authenticity.* Chicago: University of Chicago Press.

Rentfrow, P. J. and Gosling, S. D. (2003). The Do Re Mi's of Everyday Life: The Structure and Personality Correlates of Music Preferences. *Journal of Personality and Social Psychology*, **84**, 1236-1256.

Rentfrow, P. J. and Gosling, S. D. (2006). Message in a Ballad: The Role of Music Preferences in Interpersonal Perception. *Psychological Science,* **17**, 236-242.

Rentfrow, P. J. and Gosling, S. D (2007). The content and validity of music-genre stereotypes among college students. *Psychology of Music*, **35** (2), 1-21.

Scheel, K. R. and Westefeld, J. S. (1999). Heavy metal music and adolescent suicidality: An empirical investigation. *Adolescence*, **34** (134), 253-273.

Schwartz, K. D. and Fouts, G. T. (2003). Music preferences, personality style, and developmental issues of adolescents. *Journal of Youth and Adolescence*, **32** (3), 205-213.

Zweigenhaft, R. L. (2008). A Do Re Mi Encore: A Closer Look at the Personality Correlates of Music Preferences. *Journal of Individual Differences*, **29** (1), 45-55.

In: Music: Composition, Interpretation and Effects
Editor: Tamara E. Ivanova, pp. 117-126
ISBN: 978-1-60876-170-8

Chapter 7

MUSIC MEMORY AND RIGHT HIPPOCAMPUS

Takamitsu Watanabe[1,2,♦] and Hideyuki Kikyo[2,*]
[1] Department of Physiology,
the University of Tokyo School of Medicine, Tokyo, Japan
[2] Department of Molecular Neuroimaging Center,
National Institute of Radiological Sciences, Chiba, Japan

Abstract

Although it has been established that the hippocampus plays a crucial role in verbal episodic memory, its role in music memory remains unknown. Here we introduce the recent findings on the relationship between music memory and the right hippocampus. First, we introduce our previous study which demonstrated that the right hippocampus selectively responded to retrieval success of music memory. The study also demonstrated a clear functional dissociation between the left and right hippocampi in successful retrieval of music. Second, we briefly review studies that showed the recruitment of the right hippocampus related to cognitive processing of music. Many of those studies were conducted with patients suffering from brain lesions. Because there were very few cases in which the brain lesions were limited to the hippocampus, it is difficult to identify the precise role of the hippocampus. Because systematic brain imaging studies have just started, it is premature to conclude the role of the hippocampus in music processing. Third, we review studies that showed hippocampal activity related to successful retrieval of verbal memory. Some schemas have been proposed about the role of the hippocampus in memory. They are not independent of each other, and may reflect several aspects of the function of the hippocampus. The role of the hippocampus may be too complex to be explained by a single schema. We hope that our experiment of retrieval success of music will contribute to better understanding hippocampal functions.

♦ Hongo 7-3-1, Bunkyo-ku, Tokyo 113-0033, Japan.
* E-mail address: kikyo-tky@umin.ac.jp. TEL: + 81-43-206-4713, FAX: + 81-43-253-0396. Corresponding author: Hideyuki Kikyo, 4-9-1 Anagawa, Inage-ku, Chiba 263-8555, Japan.

Introduction

There are some well-known anecdotes about memory of music that many people have heard at least once. For example, young Mozart was taken by his father to hear Allegri's *Miserere*, which Rome's Vatican Choir regarded as its personal property. He went home and wrote down the whole piece from memory. Later, it was performed in the presence of a papal singer who knew the piece well, he confirmed that it was absolutely correct (Baddeley, 1998). In another example, Toscanini wanted the string section of N.B.C. Orchestra to play the slow movement of *Quartet No.5* by Joachim Raff. The score of this somewhat obscure piece could not be found in New York. Toscanini, who had probably not seen it for years, promptly wrote the whole movement down. Some time later, a collector of musical scores found a copy and detected exactly one error when he checked it against the Toscanini manuscript (Baddeley, 1998). These anecdotes are old and may have departed somewhat from the truth over time; however, such feats of 'super-memory' are not historically exceptional. Professional or talented violinists play Tchaikovsky's *Violin Concerto* lasting more than 30 min without looking at the score. They have to recall series of pitches and rhythms one after another and immediately reflect these complex musical elements in complex coordinated finger movements. Our question is whether there is specialized brain mechanism for memory of music. It is widely accepted that the hippocampus is essential for verbal episodic memory. Is the hippocampus also involved in music memory? Many people would agree that verbal information is mainly processed in the left side of the brain, and would agree that music include much non-verbal information. Which side of the brain is mainly engaged in music memory? In this report, we mainly introduce the relation between memory of music and the hippocampus.

Line of evidence shows that the hippocampus, which is one of the principal components of the medial temporal lobe, is crucial for verbal episodic memory. There are numerous studies of the hippocampus at the molecular, cellular, regional, and individual levels. Recent development of neuroimaging studies have contributed to our understanding of the mechanism of human memory at individual level. Although more than ten years have passed since functional magnetic resonance imaging (fMRI) was first used to imaging brain function, many studies of memory employed verbal stimuli. Therefore, the role of the hippocampus in memory of music, which would be non-verbal stimuli, remains poorly understood despite some evidence that brain mechanisms of music memory differ from those of verbal memory. Generally, verbal information is believed to be mainly processed in the dominant hemisphere, while non-verbal information is processed in the non-dominant hemisphere. In fact, previous anatomical studies showed that the volume of the hippocampus in the dominant hemisphere was correlated with performance of verbal memory (Richardson et al., 2004), while the volume of the non-dominant hippocampus was correlated with the amout of time spent as a taxi driver in London (Maguire et al., 2000), who would have much training in spatial recognition. Many of us would agree that music is classified as a form of non-verbal information, but it remains unknown whether this right/left hemisphere rule can apply to the processing of music memory. In order to answer the question, we first introduce our previous study, which investigated neural correlates of retrieval success of music memory and then examined the components of music the activated region responded to. Second, we briefly review studies that show the recruitment of the right hippocampus in cognitive processing of music. Finally, we review studies that showed hippocampal activity related to retrieval success of verbal memory.

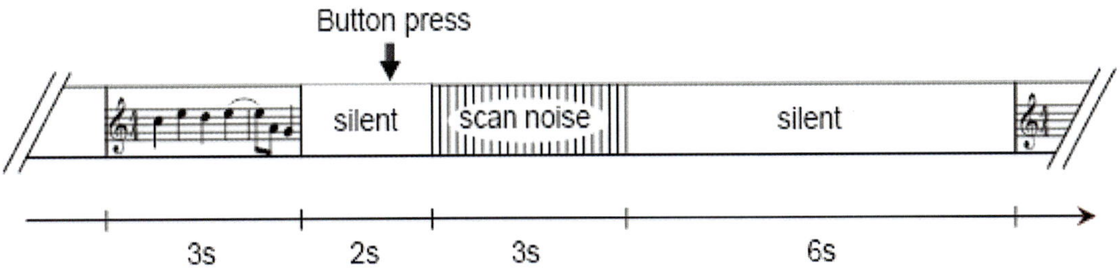

Figure 1. Sparse Temporal Sampling Technique in the Functional MRI. A newly composed music stimulus was presented for 3 s to the subjects, who were required to indicate by pressing buttons whether they recognized the stimulus as having been learned in the previous encoding phase. In the functional MRI scans, it took 3 s to scan a whole brain volume, and each volume was obtained every 14 s. For the rest of the time, there were neither sound stimuli nor MRI scan noises.

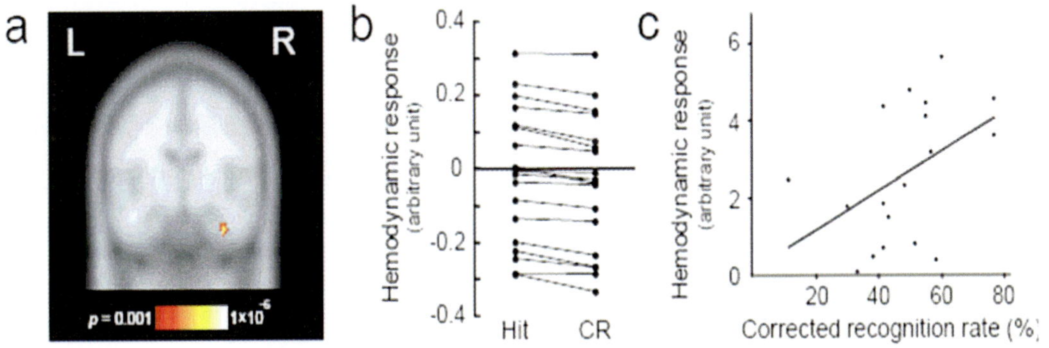

Figure 2. Retrieval Success Response in the Right Hippocampus. (a) The activation in the right hippocampus was determined by the comparison between Hit and Correct Rejection by the random effects analysis. The activation is displayed on the normalized image at the threshold of $P < 0.001$ (uncorrected). L, left; R, right; (b) The responses in Hit (left) and Correct Rejection (CR) for each subject. The activities in Hit were stronger than those in Correct Rejection in all except one subject. (c) Relationship between the retrieval success activities and the corrected recognition rates. The retrieval success activity for each subject is plotted against the corrected recognition rate. The correlation coefficient was significant ($r = 0.46$, $P = 0.05$), suggesting that the region contributes to the accuracy of music retrieval.

Retrieval Success of Music Memory

First, we investigated neural correlates of "retrieval success" for music memory using event-related functional MRI (Watanabe et al., 2008). "Retrieval success" is defined by the comparison between Hit (correctly recognized the target as having been learned in the encoding phase) and Correct Rejection (CR, correctly identified the target as not having been learned), which refers to processes engaged when a retrieval attempt is successful and yields information about a past event (Rugg et al., 1998). This is one of the basic components that constitute memory retrieval processes. Eighteen subjects participated, and none of them had received professional musical education, nor did any have any special musical preference (Wertheim and Botez's class I, Wertheim and Botez, 1959; Platel et al., 2003). None had absolute pitch. All the music stimuli were newly created for this study with auto-composing software, and all of them were in accordance with the Western musical system (Liegeois-Chauvel et al., 1998; Ayotte et al., 2000; Gaab et al., 2003; Tillmann et al., 2003). The cognitive task was the standard yes/no procedure. In the retrieval phase, the subjects were asked to listen to 30 stimuli (20 "old", 10 "new") under fMRI scannings. The sparse temporal sampling technique (Moelker and Pattynama, 2003; Nebel et al., 2005) was used to minimize possible contamination from activations induced by MRI scan noise. The musical stimulus was presented for 3 s, and the subjects were requested to press one of two buttons within 2 s, and then the fMRI scan continued for 3 s. It took 3 s to scan a whole brain volume, and each volume was obtained every 14 s (Figure 1).

In the whole-brain analyses, we found significant retrieval success activities (Hit - CR, $P< 0.001$, uncorrected) in the right hippocampus, left inferior frontal gyrus (dorsolateral prefrontal cortex, DLPFC), bilateral middle temporal gyri, left superior temporal gyrus, and left precuneus. Figure 2a shows the activation in the right hippocampus. Figure 2b shows the responses of Hit and CR for each subject, and the activities in Hit were stronger than those in CR in all except one subject. We then examined the relationship between the hemodynamic response and the correctness of the answers (corrected recognition rate) at the peak coordinates in the right hippocampus activation (Figure 2c). We found a significant correlation (r = 0.46, P = 0.05), suggesting that the right hippocampus contributes to the accuracy of music retrieval.

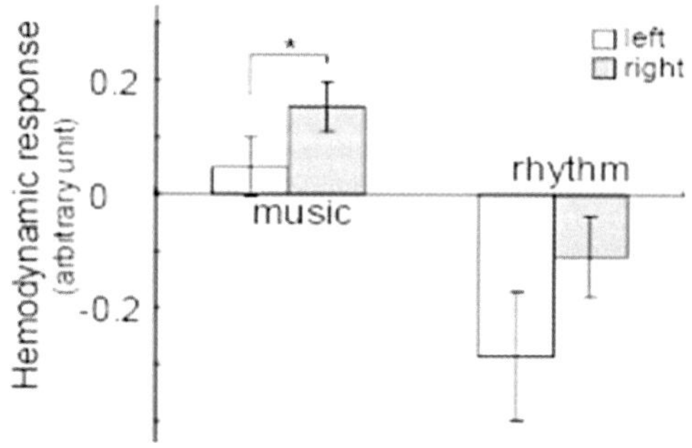

Figure 3. Differential responses of the bilateral amygdalae to music and rhythm. Those hemodynamic responses were averaged in the left and right hippocampus, which was anatomically determined Among the four conditions, only the response in the right hippocampus to the retrieval success of music was significant, and the retrieval success activity in the right hippocampus was larger than that in the left (P = 0.02). * $P< 0.05$, two-tailed, paired t-test. The error bars show standard error of mean (SEM).

Next, we investigated characteristic features of the hippocampus by region-of-interest (ROI) analysis. In the whole-brain analysis, we detected the retrieval success activation in the right hippocampus, but there is little information about the responsiveness of the left hippocampus other than the information that the retrieval success activity in the left hippocampus did not reach the significant level. The ROIs were anatomically defined, and they were delineated using MARSeille Boîte À Région d'Intérêt (MarsBaR, http://marsbar.sourceforge.net/) with Anatomical Automatic Labeling library. Using this ROI analysis, we can directly compare the activities of the left and right hippocampus. Figure 3 (left) shows the retrieval success activities in the left and right hippocampi. The response of the left hippocampus did not differ from zero ($P = 0.36$), while that of the right was significantly larger than zero ($P = 0.003$). A direct comparison between the left and right responses revealed the significantly stronger response in the right than in the left ($P = 0.02$), suggesting that there was clear functional dissociation between the bilateral hippocampi in the activities of music retrieval success.

In order to examine whether the right hippocampus was involved specifically in the music retrieval success or involved generally in a non-verbal retrieval success, we performed additional fMRI experiments with 10 subjects. The mean ages between the original and the additional experiments was not significantly different (P = 0.62). The only difference between those experiments was the to-be-remembered stimuli. In the additional experiments, a series of notes with the same pitch and varied duration (i.e. rhythm) was used. The other attributes of the stimuli, such as the sound source (piano), the length (3 s), and the sound pressure, were the same as the stimuli used in the original experiments. There was no significant difference in Hit rates between the original and the additional experiments (P = 0.50), but the corrected recognition rate of the original experiment was larger than that of the additional experiment (P < 0.01). Although, it needs careful interpretation because the number of the subjects participating in those studies was different between the original (18 subjects) and additional (10 subjects) experiments, there was no significant difference in the mean response time of Hit (P = 0.61). In the whole brain analysis of the additional experiment, the right hippocampus did not show significant activities. In the anatomically defined ROI analyses using MarsBaR, the responses of both the left and right hippocampus were not significant (Figure 3, right). Those results suggest that the right hippocampus selectively responds to the music retrieval success.

Music Related Activity in the Right Hippocampus

It is widely accepted that the hippocampus is involved in verbal episodic memory, but there is little information about the role of the right hippocampus in cognitive processing of music. Here, we review studies showing that the right hippocampus is associated with the cognitive processing of music. There are very few memory encoding studies of music. A positron emission computed tomography (PET) study contrasted the music listening state to resting state, and reported the right hippocampal activation (Brown et al., 2004). Because the comparison between passive listening and resting state would involve a cognitive component of memory encoding, the result suggests that the right hippocampus may be involved in the encoding of music.

Memory retrieval studies of music have been largely performed in subjects with brain lesions. The subjects were patients with epilepsy, which cannot be controlled by medication. To relieve epilepsy, brain resection was performed typically in the lateral temporal lobe, including the primary auditory cortex (Heschl's gyrus); the associative auditory cortex, and the medial temporal lobe. Those studies consistently showed that hippocampus was related to music retrieval (Samson and Zatorre, 1991; Samson and Zattore, 1992). Another study showed that the hippocampus was related to pleasant/unpleasant evaluations (Gosselin et al., 2006), and was related to the recognition of negative emotional music memory (Gosselin et al., 2005). A neuroimaging study showed that the chill intensity of music stimuli was negatively correlated with blood flow in the left hippocampus, but no significant change was detected in blood flow in the right hippocampus (Blood and Zatorre, 2001). Because there were very few cases in which the brain lesions were limited to the hippocampus, it is difficult to identify the precise role of the hippocampus. Because systematic brain imaging studies have just started, it is premature to conclude the role of the hippocampus in music processing.

Music can be classified as a non-verbal stimulus, but there is no evidence that we do not use language in the cognitive processing of music. It may be possible that we use language as a tool to process non-verbal information related to music. Although the previous results mentioned above are not inconsistent with our finding that the right hippocampus was recruited in the retrieval success of music memory, it remains unknown whether right hippocampal retrieval success activation reflects verbal or non-verbal information processing. The same is applicable to the prefrontal cortex. In our previous study we also found retrieval success activity in the left DLPFC, and numerous studies reported its association with verbal working memory. Our study could not discern whether the activation of the left DLPFC reflected verbal working memory or non-verbal processing which cannot be coded by language. We assume that it is useful to perform another experiment with subjects who have absolute pitch. Because music retrieval strategy can differ between people with and those without absolute pitch (Gaab *et al.* 2003), the left prefrontal region would show different responses between those with and without absolute pitch.There are many important questions to be solved in music memory, and future studies of music memory will provide further understanding of the mechanism of human memory system.

Hippocampal Activity Related to Verbal Memory Retrieval Success

A number of animal and human brain lesion studies showed that the hippocampus is essential for verbal memory. However, early neuroimaging studies detecting hippocampal activation in relation to retrieval success are limited. Before fMRI was widely used as a neuroimaging method, PET and the event-related potential were largely used to detect relationship between a particular brain structure and its cognitive function. Most PET studies used a block task design because of the intrinsic limitation of the method, which visualizes brain functions by means of radioactive tracers. These PET studies used a yes/no recognition task, but did not find hippocampal activation related to retrieval success (Nyberg et al., 1995; Rugg et al., 1996; Rugg et al., 1998). In the late 1990s, the early fMRI studies also employed a yes/no recognition task with a block design, and also failed to find hippocampal activation related to retrieval success (Buckner et al, 1998). An event-related task design become widely used in subsequent fMRI studies. However, hippocampal activation was still not observed (McDermott et al., 2000; Maratos et al., 2001; Kahn et al., 2004; Smith et al., 2004) despite an improved method of analysis (Konishi et al., 2000). Retrieval success activity in the hippocampus was also not found when photographs were used as stimuli (Weis et al., 2004). Another study using a remember/know/don't know procedure did not observe hippocampal activity related to retrieval success (Henson et al., 1999). The remember/know/don't know procedure assumes that explicit successful recognition consists of more than one underlying process, whereas the yes/no procedure assumes that such recognition consists of one process.

On the other hand, there are at least three event-related fMRI studies that reported hippocampal activity related to retrieval success. The first study used line drawings of common namable objects (Object) and their verbal names (Word) as stimuli (Stark and Squire, 2001). Left hippocampal activity related to retrieval success was observed in the Word–Word condition (names of objects were presented both at study and test), while right hippocampal activation was observed in the Object–Object (pictures of nemeable objects were presented at study and test) and Object–Word (pictures of nemeable objects were presented at study, and names of objects were presented at test) conditions. Because Object–Word condition would require associative memory, the hippocampus was considered to be no more active in the associative recognition task than in the traditional yes/no recognition task. The second study used nouns and verbs as stimuli, and administered recognition task and semantic judgment tasks within a session (Donaldson et al., 2001). They showed significant left hippocampal activity in the recognition task. The third study used a remember/know procedure, and succeeded to isolate left hippocampal activity related to explicit recollection from that related to familiarity by the comparison between remember and know condition (Eldridge et al. 2000). There may be no fundamental differences between the studies that detected hippocampal activation and those that did not, but the strength of the magnetic field of the MRI, the design of cognitive task, and the data analyses may have affected the results.

Several theories regarding the role of the hippocampus in memory have been proposed over the years. (1) Declarative memory theory (episodic memory and semantic memory) (Tulving and Donaldson, 1972): long-term memory was classified into declarative (explicit) and non-declarative (implicit) memory, and the declarative memory was subcategorized into episodic and semantic memory (Squire, 1992). The episodic memory refers to the ability to recollect individual events, and the essence of this type of memory is its specificity—its ability to represent a specific event and to locate it in time and space (Baddeley, 1998). The semantic memory reflects knowledge of the world (Baddeley 1998). The retrieval of episodic memory is dependent on the hippocampus throughout the life of a person, and becomes more resistant to partial damage with repetition and/or rehearsal. On the other hand, semantic memory becomes independent of the hippocampus and is stored in other brain regions over time (Nadel and Moscovitch, 1997; for review see Bird and Burgess, 2008). (2) Learning association and learning single items: learning association is the ability to combine two unrelated items into memory and it depends more on the hippocampal region than the ability to learn single items. This idea is plausible and there is some support for this view, but there was no suggestion that selective or disproportionate hippocampal damage impaired the ability to reject recombined stimulus elements (Kroll et al., 1996; Giovanello et al., 2003; for review, see Squire, 2004). (3) Dual process theory (recollection and familiarity): recognition memory is considered to include two components; a recollective component and a familiarity component. Recollection provides information about the episode in which an item was encountered, and familiarity provides information that an item was encountered but does not provide any knowledge about the learning context. A recent review suggested that the hippocampus is critical for recollection but not familiarity (for review, see Squire et al., 2004). The parahippocampal cortex also contributes to recollection, possibly via the representation and retrieval of contextual information, whereas the perirhinal cortex contributes to and is necessary for familiarity-based recognition (for review, see Eichenbaum et al., 2007). (4) Relational theory (Cohen and Squire, 1980; Cohen et al., 1999): The hippocampus allows the flexible association of information in neocortical modules that do not communicate otherwise. This enables the relations between elements of a scene or event to be retrieved or used for inference in novel situations, in addition to retrieval of the elements themselves. All the schemas stated above are not independent of each other, and may reflect some aspects of the function of the hippocampus, but the role of the hippocampus may be too complex to be explained by a single schema. In sum, our and the previous studies showed that there are specialized brain mechanisms in memory retrieval of music. Those studies showed that the hippocampus is included in memory retrieval of music as well as verbal materials, suggesting that there may be shared brain mechanisms between music and verbal memory retrieval. Because there were very few cases in which the brain lesions were limited to the hippocampus, it is difficult to identify the precise role of the hippocampus from lesion studies. Because systematic brain imaging studies have just started, it is premature to conclude the role of the hippocampus in music processing. Memory studies using non-verbal stimuli such as music will contribute to better understanding of human memory system.

Acknowledgment

This research was supported by the Ministry of Education, Science, Sports and Culture, Grant-in-Aid for Scientific Research (C) (19591389) to H.K.

References

Ayotte J, Peretz I, Rousseau I, Bard C, Bojanowski M (2000) Patterns of music agnosia associated with middle cerebral artery infarcts. *Brain* **123**: 1926-1938.

Baddeley A (1998) Human memory: theory and practice. MA: Allyn and Bacon.

Bird CM, Burgess N (2008) The hippocampus and memory: insights from spatial processing. *Nat. Rev. Neurosci.* **9**:182-194.

Blood AJ, Zatorre RJ (2001) Intensely pleasurable responses to music correlate with activity in brain regions implicated in reward and emotion. *Proc. Natl. Acad. Sci. USA* **98**: 11818-11823.

Brown S, Martinez MJ, Parsons LM (2004) Passive music listening spontaneously engages limbic and paralimbic systems. *Neuroreport* **15**:2033-2037.

Buckner RL, Koutstaal W, Schacter DL, Wagner AD, Rosen BR (1998) Functional-anatomic study of episodic retrieval. I. retrieval effort versus retrieval success. *NeuroImage* **7**: 151-162.

Cohen NJ, Squire LR (1980) Preserved learning and retention of pattern-analyzing skill in amnesia: dissociation of knowing how and knowing that. *Science* **210**: 207-210.

Cohen NJ, Ryan J, Hunt C, Romine L, Wszalek T, Nash C (1999) Hippocampal system and declarative (relational) memory: summarizing the data from functional neuroimaging studies. *Hippocampus* **9**: 83 - 98.

Donaldson DI, Petersen SE, Buckner RL (2001) Dissociating memory retrieval processes using fMRI: evidence that priming does not support recognition memory. *Neuron* **31**:1047-1059.

Eichenbaum H, Yonelinas AP, Ranganath C (2007) The medial temporal lobe and recognition memory. *Annu. Rev. Neurosci.* **30**:123-152.

Eldridge LL, Knowlton BJ, Furmanski CS, Bookheimer SY, Engel SA (2000) Remembering episodes: a selective role for the hippocampus during retrieval. *Nat. Neurosci.* **3**: 1149–1152.

Gaab N, Gaser C, Zaehle T, Jancke L, Schlaug G (2003) Functional anatomy of pitch memory. an fMRI study with sparse temporal sampling. *NeuroImage* **19**: 1417-1426.

Giovanello KS, Verfaellie M, Keane MM (2003) Disproportionate deficit in associative recognition relative to item recognition in global amnesia. *Cogn. Affect Behav. Neurosci.* **3**:186–194.

Gosselin N, Peretz I, Noulhiane M, Hasboun D, Beckett C, Baulac M, Samson S (2005) Impaired recognition of scary music following unilateral temporal lobe excision. *Brain* **128**:628-640.

Gosselin N, Samson S, Adolphs R, Noulhiane M, Roy M, Hasboun D, Baulac M, Peretz I (2006) Emotional responses to unpleasant music correlates with damage to the parahippocampal cortex. *Brain* **129**: 2585-2592.

Henson RN, Rugg MD, Shallice T, Josephs O, Dolan RJ (1999) Recollection and familiarity in recognition memory: an event-related functional magnetic resonance imaging study. *J. Neurosci.* **19**: 3962-3972.

Kahn I, Davachi L, Wagner AD (2004) Functional-neuroanatomic correlates of recollection: implications for models of recognition memory. *J. Neurosci.* **24**:4172-4180.

Konishi S, Wheeler ME, Donaldson DI, Buckner RL (2000) Neural correlates of episodic retrieval success. *NeuroImage* **12**: 276-286.

Kroll NE, Knight RT, Metcalfe J,Wolf ES, Tulving E (1996) Cohesion failure as a source of memory illusions. *J. Mem. Lang.* **35**:176–196.

Liegeois-Chauvel C, Peretz I, Babai M, Laguitton V, Chauvel P (1998) Contribution of different cortical areas in the temporal lobes to music processing. *Brain* **121**: 1853-1867.

Maratos EJ, Dolan RJ, Morris JS, Henson RN, Rugg MD (2001) Neural activity associated with episodic memory for emotional context. *Neuropsychologia* **39**:910-920.

Maguire EA, Gadian DG, Johnsrude IS, Good CD, Ashburner J, Frackowiak RS, Frith CD (2000) Navigation-related structural change in the hippocampi of taxi drivers. *Proc. Natl. Acad. Sci. USA* **97**:4398-4403.

McDermott KB, Jones TC, Petersen SE, Lageman SK, Roediger HL 3rd (2000) Retrieval success is accompanied by enhanced activation in anterior prefrontal cortex during recognition memory: an event-related fMRI study. *J. Cogn. Neurosci.* **12**: 965-976.

Moelker A, Pattynama PM (2003) Acoustic noise concerns in functional magnetic resonance imaging. *Hum. Brain Mapp.* **20**: 123-141.

Memory consolidation, retrograde amnesia and the hippocampal complex (1997) Nadel L, Moscovitch M. *Curr. Opin. Neurobiol.* **7**:217-227.

Nadel L, Moscovitch M (1997) Memory consolidation, retrograde amnesia and the hippocampal complex. *Curr. Opin. Neurobiol.* **7**:217-227.

Nebel K, Stude P, Wiese H, Muller B, de Greiff A, Forsting M, Diener HC, Keidel M (2005) Sparse imaging and continuous event-related fMRI in the visual domain: a systematic comparison. *Hum. Brain Mapp.* **24**: 130-143.

Nyberg L, Tulving E, Habib R, Nilsson LG, Kapur S, Houle S, Cabeza R, McIntosh AR (1995) Functional brain maps of retrieval mode and recovery of episodic information. *Neuroreport* **7**: 249-252.

Platel H, Baron JC, Desgranges B, Bernard F, Eustache F (2003) Semantic and episodic memory of music are subserved by distinct neural networks. *NeuroImage* **20**: 244-256.

Richardson MP, Strange BA, Thompson PJ, Baxendale SA, Duncan JS, Dolan RJ (2004) Pre-operative verbal memory fMRI predicts post-operative memory decline after left temporal lobe resection. *Brain* 127: 2419-2426.

Rugg MD, Fletcher PC, Frith CD, Frackowick RSJ, Dolan RJ (1996) Differential activation of the prefrontal cortex in successful and unsuccessful memory retrieval. *Brain* **119**: 2073-2083.

Rugg MD, Fletcher PC, Allan K, Frith CD, Frackowiak RS, Dolan RJ (1998) Neural correlates of memory retrieval during recognition memory and cued recall. *NeuroImage* **8**: 262-273.

Samson S, Zatorre RJ. (1991) Recognition memory for text and melody of songs after unilateral temporal lobe lesion: evidence for dual encoding. *J. Exp. Psychol. Learn Mem. Cogn.* **17**:793-804.

Samson S, Zatorre RJ (1992) Learning and retention of melodic and verbal information after unilateral temporal lobectomy. *Neuropsychologia*: 815-826.

Smith AP, Henson RN, Dolan RJ, Rugg MD (2004) fMRI correlates of the episodic retrieval of emotional contexts. *Neuroimage* **22**:868-878.

Squire LR (1992) Declarative and non-declarative memory: multiple brain systems supporting learning and memory. *J. Cognit. Neurosci.* **4**: 232-243.

Squire LR, Stark CE, Clark RE. The medial temporal lobe (2004) *Annu. Rev. Neurosci.* **27**: 279-306.

Stark CE, Squire LR (2001) Simple and associative recognition memory in the hippocampal region. *Learn Mem.* **8**:190-197.

Tillmann B, Bharucha JJ, Bigand E (2003) Learning and perceiving musical structures: further insights from artificial neural network. In: The Cognitive Neuroscience of Music (Peretz I, Zatorre RJ, ed). New York, Oxford University Press.

Tulving E, Donaldson W (1972) Organization of memory. New York: Academic Press.

Watanabe T, Yagishita S, Kikyo H (2008) Memory of music: roles of right hippocampus and left inferior frontal gyrus. *NeuroImage* **39**:483-491.

Weis S, Klaver P, Reul J, Elger CE, Fernández G (2004) Temporal and cerebellar brain regions that support both declarative memory formation and retrieval. *Cereb. Cortex.* **14**: 256-267.

Wertheim N, Botez MI (1959) Plan for the examination of musical functions. *Encephale* **48**: 246-254.

In: Music: Composition, Interpretation and Effects
Editor: Tamara E. Ivanova, pp. 127-132
ISBN: 978-1-60876-170-8

Chapter 8

Cerebral Areas Involved in Music Perception: Studying Musicogenic Epilepsy

Francesca Pittau[1], Francesca Bisulli[1], Carlotta Stipa[1], Ilaria Naldi[1], Laura Licchetta[1], Lidia Di Vito[1], Daniela Cevolani[2], Raffaele Agati[2] and Paolo Tinuper[1]

[1] Department of Neurological Sciences,
University of Bologna, Bologna, Italy
[2] Neuroradiology Unit, Department of Neurosciences,
Ospedale Bellaria, Bologna, Italy

Abstract

Functional cerebral areas involved in listening to emotionally charged music are wider than those activated by neutral music. Based on the pathophysiology of musicogenic epilepsy, a rare form of complex reflex epilepsy in which seizures are triggered by music, we can infer the functional organization of the brain's processing of music. We studied a 36-year-old, right-handed male amateur musician, who has had weekly epileptic partial seizures every time he listened to or played music with a strong emotional charge since the age of 24 years. The patient underwent prolonged video-polygraphic recording which documented three right temporal seizures preceded for several seconds by an increase in heart rate and blood pressure, all triggered by listening to music with an emotional content. During functional neuroimaging (fMRI) he listened to both "neutral" and "emotionally charged" music. The "neutral music" activated only a small eloquent area in the right temporal lobe (acoustic area), whereas an "emotionally charged melody" activated diffuse eloquent areas on the fronto-temporo-occipital lobes of the right hemisphere before seizure onset. Numerous studies have demonstrated the predominant involvement of right hemisphere structures in networks involved in processing musical information. Most cases of musicogenic epilepsy showed the dominant role of the right temporal lobe disclosed by functional neuroimaging and neurophysiological techniques. The triggering stimulus in our patient seems to be a strong emotional feeling induced by specific melodies, as his seizures were preceded by an increase in heart rate and blood pressure. Our results emphasize the role of the right temporal lobe in musicogenic epilepsy and show that the cerebral areas activated during the emotional status leading to seizures encompass the auditory cortex activated by neutral music.

Introduction

Epilepsy is a neurological disorder characterized by recurrent epileptic seizures. The seizures are transient symptoms and/or signs of abnormal and synchronous neuronal activity in the brain. The prevalence of epilepsy is roughly 0.5-1 per 100 people, while the annual incidence rate is 29-53 per 100.000 [1]. Epileptic seizures can be generalized or focal and they tend to occur spontaneously, but in rare cases can be triggered by specific stimuli (reflex epilepsy). At present, reflex epilepsy is defined in the Engel proposed classification as a specific syndrome "in which all epileptic seizures are precipitated by sensory stimuli" [2] thereby excluding the reflex seizures that occur in focal and generalized conditions that are also associated with spontaneous seizures. The stimuli precipitating reflex seizures include: visual, auditory, thinking, praxis, somatosensory and proprioceptive stimuli, reading, eating, and hot water immersion. The 1985 classification [3] divides reflex epilepsies into "simple" and "complex". Simple reflex epileptic seizures are provoked by a simple sensory stimulus, usually unexpected or of high intensity, with a short latency between the stimulus and the seizure. Complex reflex epileptic seizures have been characterized as being triggered by elaborate stimuli in which a specific pattern, sometimes just the expectation of the provoking stimulus, is the triggering factor, suggesting the involvement of higher cortical functions. The interval from the stimulus to the beginning of the seizure is normally longer than in simple reflex seizures. Musicogenic epilepsy is characterized by seizures induced typically by listening to music [4]. Although some cases had been reported in the past, Critchley was the first to define this syndrome [5]. Musicogenic epilepsy is a uncommon form of complex reflex epilepsy in which seizures are triggered by music ("sounds in melodic or harmonic combination") [6]. Musicogenic seizures are usually focal and involve the temporal regions with a right-sided prevalence [4,7]. They may be variably followed by loss of consciousness and, more seldom, by secondarily generalized seizures. Musicogenic seizures are typically provoked by a specific musical stimulus differing in each patient. Therefore, more knowledge of pathophysiological features of this kind of epilepsy would shed more light on the neuronal networks underlying music processing in the brain [8].

Although listening to music is the most common precipitating factor, thinking, playing or dreaming about music have also been reported [9-12]. An affective or emotional content of the stimulus is often described. We studied [13] a 36-year-old, right-handed male amateur musician (he used to play classical guitar in a folk band). Since the age of 24 years, he has had weekly epileptic seizures, lasting about 20 seconds, characterized by unpleasant ascending gastric sensation, tachycardia, and oroalimentary automatisms. All of his seizures were triggered by music. In particular, the patient realized that playing the guitar himself or listening to (but also thinking or dreaming about) certain pieces of music with a strong emotional content would trigger a seizure, whereas neutral songs (i.e., songs without emotional components) were ineffective. At age 27 he had a secondarily generalized seizure. Since the age of 27, seizures began with fear and an unpleasant ascending gastric sensation, and he has avoided listening to or playing music. Antiepileptic drugs were ineffective. The patient agreed to undergo prolonged videopolygraphic recording during which he listened to both neutral music ("La Canzone del Sole," played by a popular Italian folk singer) and emotionally charged music with a very strong emotional content for the patient (the song "Overdue Goodbye" played by a famous American rock star). To explore the autonomic nervous system changes associated with the seizures, together with EEG we monitored the electrocardiogram, plethysmogram, thoracic breathing, and blood pressure. Listening to "neutral" music did not provoke any seizure or pathological changes in the polygraphic recording. Three typical seizures were recorded, all triggered by listening to the affective music (the first after 98 seconds, the second after 120 seconds, and the third after 113 seconds of listening to the melody). The seizures were stereotyped: the patient was able to push a button and speak, could change his facial expression, had involuntary chewing and swallowing automatisms, and signaled paresthesias in his hands. On the EEG, the seizures were accompanied by ictal paroxysmal discharges over the right temporal regions, identical in the three seizures. His cardiac rhythm and respiration rate increased slowly from the beginning of the song, reached a maximum 10 seconds before the seizure, and decreased slightly in the last part of the episode. The patient underwent brain magnetic resonance scan (MRI) and functional neuroimaging (fMRI). The fMRI block paradigm was implemented as follows: 60 seconds of silence, 45 seconds of neutral music ("La Canzone del Sole"), 60 seconds of silence, 45 seconds of epileptogenic music ("Overdue Goodbye"), followed by a 3-minute pause for autonomic values to return to normal. This procedure was repeated three times for a total duration of 210 seconds. This made the length of the epileptogenic music inadequate to produce an attack, but at the same time we obtained sufficient data for statistical analysis.

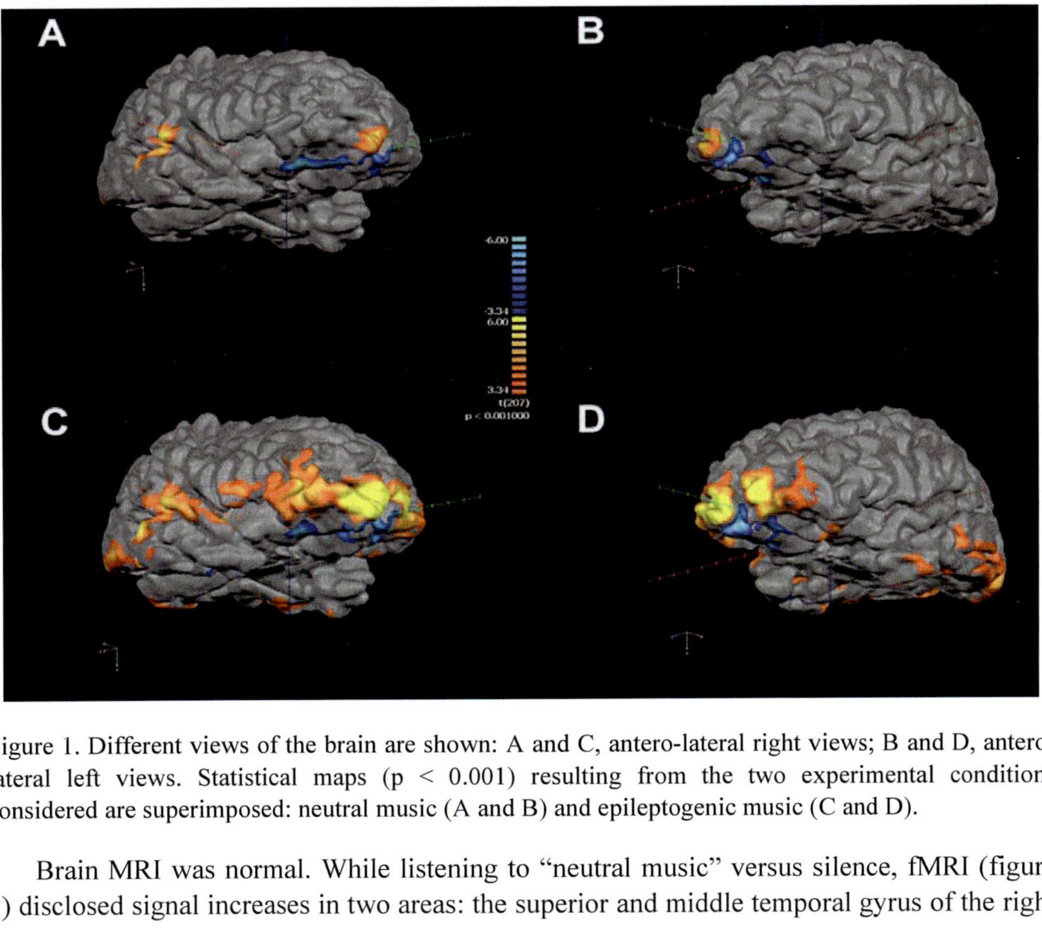

Figure 1. Different views of the brain are shown: A and C, antero-lateral right views; B and D, antero-lateral left views. Statistical maps ($p < 0.001$) resulting from the two experimental conditions considered are superimposed: neutral music (A and B) and epileptogenic music (C and D).

Brain MRI was normal. While listening to "neutral music" versus silence, fMRI (figure 1) disclosed signal increases in two areas: the superior and middle temporal gyrus of the right temporal lobe, corresponding to the acoustic area, and the right frontal pole. Other areas showed signal decreases: the right gyrus rectus, the right and left subcallosal gyrus, and small bilateral areas in the orbital gyrus. Listening to "epileptogenic music" versus silence elicited more eloquent areas, widespread mainly on the right hemisphere. Signal increases involved the right temporal areas corresponding to the acoustic areas, the posterior part of the temporal lobe, the occipital pole on the right hemisphere, the anterior frontal poles bilaterally, and the inferior surface of the brain bilaterally. The eloquent areas characterized by signal decreases were the posterior part of the right gyrus rectus, the subcallosal gyrus bilaterally, and small areas in the orbital gyrus bilaterally.

Historically, the first report a form of epilepsy related to music dates back to the XVI century when J.J. Scalinger described seizures provoked by the sound of the lira [14]. In 1847 the chinese poet Kung Tsu Chen wrote: "Since my early youth I have always been distracted when I listen to the sound of a flute hawker. I feel ill when I listen to the sound of a flute in the evening sun, although I do not know the reason" [15]. Shakespeare too may have had knowledge of similar cases, because in the Merchant of Venice (Act IV, scene I) he wrote: "Some men there are love not a gaping pig; some that are mad if they behold a cat; and others, when the bagpipe sings i' the nose, cannot contain their urine".

We performed a literature search which disclosed one hundred and ten cases of musicogenic epilepsy [13]. The first case of epilepsy provoked by music was reported in 1884 by Merzheevski [16]. He described a patient who had seizures provoked by listening to unfamiliar sounds, that ceased to cause attacks when the patient became accustomed to them. The first to coin the definition "musicogenic epilepsy" was Critchley in 1937 describing eleven patients with seizures induced by music [5]. With regard to the musical stimulus, our literature search disclosed that in some patients seizures are always triggered by the same song, for example a patient described by Lwoff whose seizures were caused only by listening to the Marseillaise [16]. In other cases seizures are always provoked by a particular kind of music: in the patient described by Bickford they were induced only by exotic jazz music [17]. In some patients it is important that the music is played by a particular instrument, for example the piano [5]. Some patients had seizures only if they were performing the music as reported by Critchley who described a patient who had seizures while he played piano [5], or by Sutherling who reported a case whose seizures were caused by playing a particular hymn on the organ [9]. Another unusual case was described by Piotrowski who wrote about a man who had seizures provoked uniquely by songs sung by female artists [18]. In many cases the stimulus is not so specific, and all kinds of music can cause a seizure [19,20]. Although listening to music is the commonest stimulus in literature reports, thinking [21] or dreaming [10] about a particular piece of music have also been noted. Moreocver, many reports emphasize the relevance of the emotional component in musicogenic seizures. Some patients referred that the musical stimulus had to be sentimental or romantic [22] whereas others noted that sadness favoured the onset of seizures [23].

Conclusion

The pathophysiology of musicogenic epilepsy remains unsettled. Melody, harmony, rhythm, emotional effect and memory are fundamental components of a musical stimulus. Each of these components could be epileptogenic for each patient [24], but it is often difficult to isolate the actual stimulus. A strong emotional feeling related to specific melodies seems to be the effective stimulus in our patient as an increase in heart rate and blood pressure precedes his seizures by several seconds. Studies in subjects with unilateral cortical lesions [25,26] and many functional neuroimaging studies [27,28] have demonstrated the predominant involvement of right hemisphere structures in networks involved in processing musical information. Zifkin and Zatorre [6] also noted that complex musical processing tasks activate bilateral cortical and subcortical territories, with right hemisphere prevalence. Our literature review disclosed 110 published cases of musicogenic epilepsy. A dominant role of the right temporal lobe has been implicated in this kind of epilepsy: right temporal predominance and right anterior and mesial hyperperfusion have been demonstrated on ictal SPECT [7,29-31]; right temporal involvement has also been shown on subdural EEGs [24]. Nevertheless other studies failed to confirm these findings, emphasizing the role of the left temporal lobe [32,33]. These discrepancies could be associated with the fact that the localizing factor is related to the musical component implicated in triggering the seizure [24]. Our results support the role of the right temporal lobe in musicogenic epilepsy and demonstrate that the cerebral areas activated during the emotional status leading to the seizures encompass the auditory cortex activated by neutral music.

References

[1] Hauser WA. Incidence and prevalence. In: Engel J Jr, Pedley TA, editors. *Epilepsy: A Comprehensive Textbook.* Philadelphia: Lippincott–Raven; 1997; 47-57.

[2] Engel, J Jr. A proposed diagnostic scheme for people with epileptic seizures and epilepsy: report of the ILAE Task Force on Classification and Terminology. *Epilepsia,* 2001 42, 796–803.

[3] Commission on Classifications and Terminology of ILAE. Proposal for revised classification of epilepsies and epileptic syndromes. *Epilepsia,* 1985 26, 268–278.

[4] Critchley, M. Musicogenic epilepsy. In: Critchley M, Henson RA, editors. *Music and the brain: Studies in the neurology of music.* London: Heinemann; 1977; 344–53.

[5] Critchley, M. Musicogenic epilepsy. *Brain,* 1937 60, 13–27.

[6] Zifkin, BG; Zatorre, RJ. Musicogenic epilepsy. In: Zifkin BG, Andermann F, Beaumanoir A, Rowan AJ, editors. *Reflex epilepsies and reflex seizures.* Advances in Neurology, vol. 75. Philadelphia: Lippincott–Raven; 1998; 273–81.

[7] Wieser, HG; Hungerbühler, H; Siegel, AM; Buck, A. Musicogenic epilepsy: review of the literature and case report with ictal single photon emission computed tomography. *Epilepsia,* 1997 38, 200–7.

[8] Avanzini, G. Musicogenic seizures. *Ann. NY Acad. Sci,* 2003 999, 95–102.

[9] Sutherling, WW; Hershman, LM; Miller, JQ; Lee, SI. Seizures induced by playing music. *Neurology,* 1980 30, 1001–4.

[10] Shaw, D; Hill, D. A case of musicogenic epilepsy. *J. Neurol. Neurosurg. Psychiatry,* 1947 10, 107–17.

[11] Stubbe-Teglbjaerg, HP. On musicogenic epilepsy. *Acta Psychiatr. Neurol,* 1949 24, 679–88.

[12] Dearman, HB; Smith, BM. A case of musicogenic epilepsy. *JAMA,* 1965 193, 1123–5.

[13] Pittau, F; Tinuper, P; Bisulli, F; Naldi, I; Cortelli, P; Bisulli, A; Stipa, C; Cevolani, D; Agati, R; Leonardi, M; Baruzzi, A. Videopolygraphic and functional MRI study of musicogenic epilepsy. A case report and literature review. *Epilepsy Behav,* 2006 13, 685-92.
[14] Kaplan, PW. Musicogenic epilepsy and epileptic music: a seizure's song. *Epilepsy Behav,* 2003 4, 464-73.
[15] Fujinawa, A; Kawai, I; Ohashi, H; Kimura, S. A case of musicogenic epilepsy. *Folia Psychiatr Neurol. Jpn,* 1977 31, 463-72.
[16] Vanelle, JM. Contribution a l'étude de l'épilepsie musicogénique. Analyse de deux observations. These, Paris University, 1982.
[17] Bickford, RG. Musicogenic epilepsy. *Electroenceph. Clin. Neurophysiol,* 1956 8, 152-3.
[18] Piotrowski, A. A case of musicogenic epilepsy. Clinical and EEG study. *Electroenceph. Clin. Neurophysiol,* 1959 11, 176.
[19] Vercelletto, P. A propos d'un cas d'épilepsie musicogénique. Présentation d'une crise temporale, discussion sur son point de départ. *Rev. Neurol,* 1953 88, 379-82.
[20] Smeijsters, H; Van den Berk, P. Music therapy with a client suffering from musicogenic epilepsy: a naturalistic qualitative single-case research. *The Arts in Psychotherapy,* 1995 22, 249-63.
[21] Nikitin, MP. Zur Psychogenese der epileptischen Anfalle. *Nervenarzt,* 1935 8, 66-9.
[22] Gastaut, H; Tassinari, CA. Triggering mechanisms in epilepsy. The electroclinical point of view. *Epilepsia,* 1966 7, 85-138.
[23] Daly, DD; Barry, MJ. Musicogenic epilepsy: report of three cases. *Psychosom. Med,* 1957 19, 399-408.
[24] Tayah, TF; Abou-Khalil, B; Gilliam, FG; Knowlton, RC; Wushensky, CA; Gallagher, MJ. Musicogenic seizures can arise from multiple temporal lobe foci: intracranial EEG analyses of three patients. *Epilepsia,* 2006 47, 1402–6.
[25] Johnsrude, IS; Penhune, VB; Zatorre, RJ. Functional specificity in the right human auditory cortex for perceiving pitch direction. *Brain,* 2000 123, 155–63. 394.
[26] Samson, S; Zatorre, RJ; Ramsay, JO. Deficits of musical timbre perception after unilateral temporal-lobe lesion revealed with multidimensional scaling. *Brain,* 2002 125, 511–23.
[27] Zatorre, RJ; Bouffard, M; Ahad, P; Belin, P. Where is ''where" in the human auditory cortex? *Nat. Neurosci*, 2002 5, 905–9.
[28] Zatorre RJ, Bouffard M, Belin P. Sensitivity to auditory object features in human temporal neocortex. *J. Neurosci,* 2004 24, 3637–42.
[29] Genç, BO; Genç, E; Taştekin, G; Iihan, N. Musicogenic epilepsy with ictal single photon emission computed tomography (SPECT): could these cases contribute to our knowledge of music processing? *Eur. J. Neurol,* 2001 8, 191–4.
[30] Gelisse, P; Thomas, P; Padovani, R; Hasson-Sebbag, N; Pasquier, J; Genton, P. Ictal SPECT in a case of pure musicogenic epilepsy. *Epileptic. Disord,* 2003 5, 133–7.
[31] Cho, JW; Seo, DW; Joo, EY; Tae, WS; Lee, J; Hong, SB. Neural correlates of musicogenic epilepsy: SISCOM and FDG-PET. *Epilepsy Res,* 2007 77, 169–73.
[32] Lin, KL; Wang, HS; Kao, PF. A young infant with musicogenic epilepsy. *Pediatr Neurol,* 2003 28, 379–81.
[33] Anneken, K; Fischera, M; Kolska, S; Evers, S. An unusual case of musicogenic epilepsy in a patient with a left fronto-temporal tumour. *J. Neurol,* 2006 253, 1502–4.

In: Music: Composition, Interpretation and Effects
Editor: Tamara E. Ivanova, pp. 133-158
ISBN 978-1-60876-170-8

Chapter 9

MUSICAL SOURCE SEPARATION: METHODS AND APPLICATIONS

Maximo Cobos and Jose J. Lopez
Institute for Telecommunications and Multimedia Applications (iTEAM),
Universidad Politecnica de Valencia, Valencia, Spain

Abstract

Source Separation methods applied to musical signals have been shown to provide very interesting results for music processing. In this context, music unmixing, automatic music transcription and source-oriented reprocessing is possible using advanced digital signal processing techniques. Most musical signals are mixtures of several sources (mainly instruments and singers) which are active simultaneously. Sound Source Separation (SSS) aims at recovering each source signal from a set of mixture signals of the original sources, such as those obtained by a stereo microphone pair or an audio CD. Although some methods are able to provide a solution to the SSS problem, they are limited to certain mixing scenarios or the resulting separation quality is not as good as expected, which may difficult their application to practical music processing systems. In fact, the objective of SSS algorithms is usually to separate with the highest quality, but it is also possible to perform separation just to facilitate a posterior semantic analysis of complex audio material. Thus, the objective of SSS methods may be completely different depending on the application. In this chapter, we review some of the existing techniques for sound source separation and we discuss their practical use for musical applications.

1. Introduction

Humans are surrounded by sound. If we try to concentrate and listen carefully to the things that happen in our environment, probably we will be able to identify more than one source of sound. This fact reveals an important property of human hearing: the ability to distinguish individual sound sources from complex mixtures of sound. When talking about the perception of speech in complex acoustic environments, this human ability is usually related to the well known "cocktail party problem" [1]. Thus, the cocktail party effect describes the ability to focus one's listening attention on a single talker among a mixture of conversations and background noises, ignoring other conversations and enabling us to talk in a noisy place. In addition to speech understanding, this special ability plays also a major role in the way humans perceive and feel music. Most of musical compositions are mixtures of different instruments playing simultaneously and we are usually able to concentrate on hearing one of these instruments. A trained subject can even describe the melodies or musical lines played by each instrument and translate them into musical language and notation. Therefore, we can generally talk about an special ability of understanding sound scenes, defined as a complex entity of acoustic stimuli that is processed by our brain, resulting in a symbolic description corresponding to a listener's perception of the different auditory events or sources that are present [2].

The complex understanding of sound scenes is so familiar to humans that we usually take it for granted and we perform this task without being aware of it. However, computational methods aimed at imitating our hearing sense and brain for understanding complex sound material are far from performing with such an accuracy and are still an active research line that involves both psychological and physical issues. Designing machines capable of hearing the way humans do is the aim of "Computational Auditory Scene Analysis" (CASA) [3], which is nowadays a hot research topic. The processes underlying the perceptual interpretation of sound sources have been studied for decades, being the work by Bregman [4] one of the most well-known references in this field. These studies on human sound perception have revealed useful information on how sound is processed in the inner ear, leading to engineering prototypes and algorithms that can, with relatively good accuracy, perform tasks like segregation of speech and musical instruments, automatic music transcription or estimation of multiple pitches from noisy mixtures. In this context, a common approach to achieve understanding of complex sound scenes is usually sound source separation (SSS). Generally, the objective of SSS methods is the extraction of each individual audio signal that constitutes a sound mixture, as depicted in Figure 1. However, it is important to notice that human understanding of sound scenes is related to the perception of some properties that suggests the existence of certain auditory events from which they can obtain a description of what they are listening. Therefore, to date, there is no necessary evidence for the human auditory system to extract the individual audio signal corresponding to each auditory event [4]. Moreover, it is not clear that sound source separation is necessary for understanding a sound scene and it could be possible to follow an *understanding without separation* approach [5]. Nevertheless, having estimations of the sources that conform a sound mixture is surely a help for analyzing the different features corresponding to different auditory events, and therefore, for providing a description of a sound scene.

Existent SSS approaches to achieve separation of musical sound scenes are very varied. In addition to CASA methods, the field of sound source separation is related to other well-known approaches in the field of statistical signal processing and machine learning. These methods are usually called Blind Audio Source Separation (BASS) methods, where the term *blind* is used to emphasize that very little information about the sources or the mixing process is known. Generally, several observations of the mixture are available in these kind of methods, but their performance is closely related to the mixing environment. Popular approaches as *Independent Component Analysis* techniques belong to this type of algorithms, where statistical assumptions about the sources are taken [6]. When there is only one observation channel available, it is necessary to assume some information about the structure of the source that has to be separated, and therefore model-based methods that take into

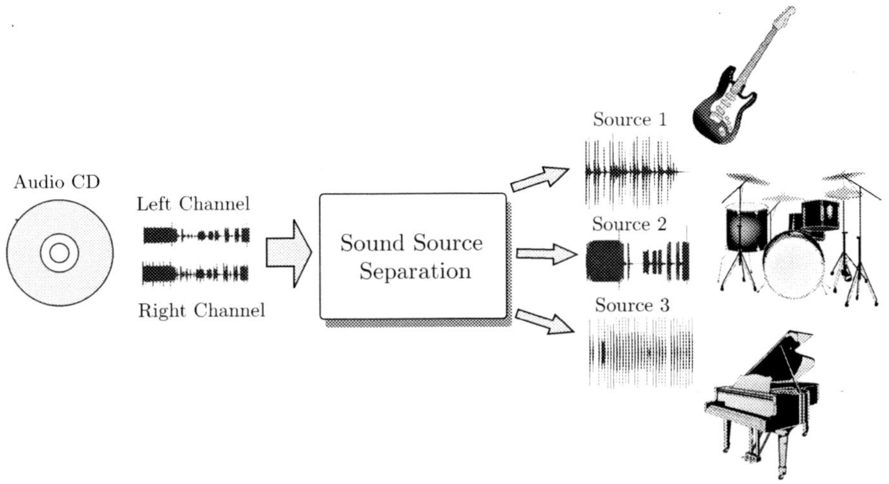

Figure 1. Sound Source Separation algorithms are aimed at extracting the different sound sources that are present in a mixture.

account the harmonic spectral structure of pitched musical instruments are used [7]. Algorithms for learning the underlying spectral structure of musical sources can also be found in the literature, such as the *Non-Negative Matrix Factorization* algorithm [8]. Especially in the case of musical sound scenes, the mixtures are usually underdetermined, i.e. there are more sources than observation channels, (think for example of an stereo CD) and *sparse* methods are used [9]. The sparsity of audio sources under a time-frequency transformation has led to popular approaches, especially those based on *time-frequency masking*.

Applications of sound source separation algorithms are often classified into two categories depending on which is the aim of the separated sources. If the separated sources are intended to be listened to, we talk about Audio Quality Oriented (AQO) applications. Otherwise, we talk about Significance Oriented (SO) applications [10].

AQO applications extract sources that are listened to, directly after separation or after some post-processing audio treatment. Therefore, they are intended to fully unmix the mixture at the highest possible quality. Some applications inside this category are:

- *Music up-mixing*: High quality audio separation would make possible to adapt mono and stereo material to present and future spatial sound systems, such as 5.1, Wave-Field Synthesis (WFS) [11] or Ambisonics [12].

- *Music remixing and reprocessing*: Extracting a sound source from a mixture can be used for karaoke applications [13] or intelligent/selective reprocessing of music.

- *Speech enhancement*: Separating concurrent speech sources can be used for improving intelligibility in speech communication systems, such as videoconference, hearing-aids or free-hand devices [14].

On the other hand, SO applications are aimed at obtaining semantic information from complex audio signals. These applications do not need to separate with excellent quality the sources, as they should only be capable of extracting relevant information from the mixture audio signal. Applications within this category include:

- *Music Information Retrieval* (MIR): This term involves all kind of techniques aimed at retrieving information from music [15], including semantic retrieval, genre classification, query by humming, etc.

- *Automatic Transcription of Polyphonic Music*: Musical transcription of audio data is the process of taking a sequence of digital data corresponding to the sound waveform and extracting from it the symbolic information related to the high-level musical structures that might be seen on a score [16]. These techniques are also related to MIR.

- *Other Applications*: Generally, carrying out a separation of the sources as a previous step for other applications can improve the performance of other systems, such as automatic speech recognition (ASR) systems [17]. Also, the information extracted from the sources can be used for the efficient coding of audio objects [18].

In this chapter, we will review some of the existing techniques for sound source separation and we will discuss their practical use for musical applications. The chapter is structured as follows. In Section 2., the source separation problem is presented, with a description of typical audio sources and mixing schemes. Section 3. is focused on the stereo underdetermined problem, explaining briefly the need for sparse transformations and the popular DUET algorithm based on time-frequency masking. Section 4. provides an overview of two common approaches for monaural separation: Non-negative Matrix Factorization and Computational Auditory Scene Analysis. In Section 5., common applications for music are described, including music up-mixing, selective reprocessing of audio objects and automatic music transcription. Finally, Section 6. summarizes the conclusions of this chapter.

2. Sound Source Mixing and Separation Tasks

2.1. Speech and Music Sources

The temporal and spectral features of audio sources are essential for many separation algorithms. Speech sources can be thought as a sequence of discrete units, called phonemes. Speech signals are non-stationary and have a characteristic structure that includes a periodic part containing harmonic sinusoidal partials and a transient or noisy part. Sinusoidal partials are multiples of a single frequency called the fundamental frequency, or pitch. The pitch varies over time, usually centered at an average frequency of 140 Hz with variations of 140 Hz for males and 200 Hz for females. Phonemes define the spectral envelope and vary among speakers and the phonetic context [19].

Musical instruments and singers produce sequences of notes. The signals produced by notes also follow a basic structure, where a transient part is usually followed by a near-periodic part of harmonic sinusoidal partials. Fundamental frequencies in music vary slower in time than speech does, and usually on discrete steps of the semitone scale, which spans logarithmically the range from 30 Hz to 4 kHz. The timbre of instruments depends on several factors, such as the duration of onset transients, the spectral envelope and its frequency modulation. As western harmony rules are based on specific frequency ratios (2/1, 3/2 or 5/4), there is usually overlapping of frequencies when musical instruments play jointly.

2.2. The Mixing Process

Mixtures of audio sources can be acquired in many ways. In this context, synthetic audio mixtures are very different in nature from real live recordings, where the mixtures are obtained by capturing several sources that have been physically mixed. Thus, the spatial properties of the signals highly depend on their mixing process. The recording set-up has a big influence on the different separation approaches. In real recordings, microphone arrangements may involve near-field and far-field microphones with certain directional patterns. Recordings by means of dummy heads are also of interest for hearing-aid applications. The room where the recording is made plays also an important role, as the recorded signals are the total contribution of direct path signals from the sources and the reflections that occur inside the room. Therefore, the observation mixtures are filtered versions of the sources due to successive reflections on the room surfaces. These reflections can be divided into early reflections and late reflections. Early reflections may be calculated easily from the room geometry, but late reflections make off the room reverberation, which has an stochastic structure. Reverberation is usually characterized by the room reverberation time, RT_{60}, which is the time required for reflections of a direct sound to decay by 60 dB below the level of the direct sound. Reverberation time is defined for wide band signals. When talking about the decay of an individual frequency, the term decay time is used. Normal RT_{60} values range from 100 ms to 500 ms in office rooms and more than 1 s in large venues.

Music productions from studios (such as pop music or movie soundtracks) are often made by recording separately the sources in a near-anechoic room with a single microphone and applying posterior effects to each source. Common effects include panoramic and reverberation. The panoramic pot modifies the spatial location of the sources in the mix by applying a different factor scale in each channel. Reverberation simulates the effect introduced by a room by applying filters to the sources. In the down-mix step, all the sources are added synthetically by means of an audio workstation or a mixer. Synthetic mixing can also be used to add the channels of a live recording. Figure 2 shows several mixing set-ups classified according to their type of mixing. In the following subsections we describe how these situations are mathematically modeled.

2.3. Mixing Models

We consider a general set-up where the number of sources is N and M observation channels are available. Generally, the mixing process can be written as

$$x_m(t) = \sum_{n=1}^{N} s_{mn}(t), \quad m = 1, \ldots, M, \tag{1}$$

where s_{mn} is the image of the n-th source in the m-th microphone. These images of the sources represent how the original source signals are recorded at each sensor after being

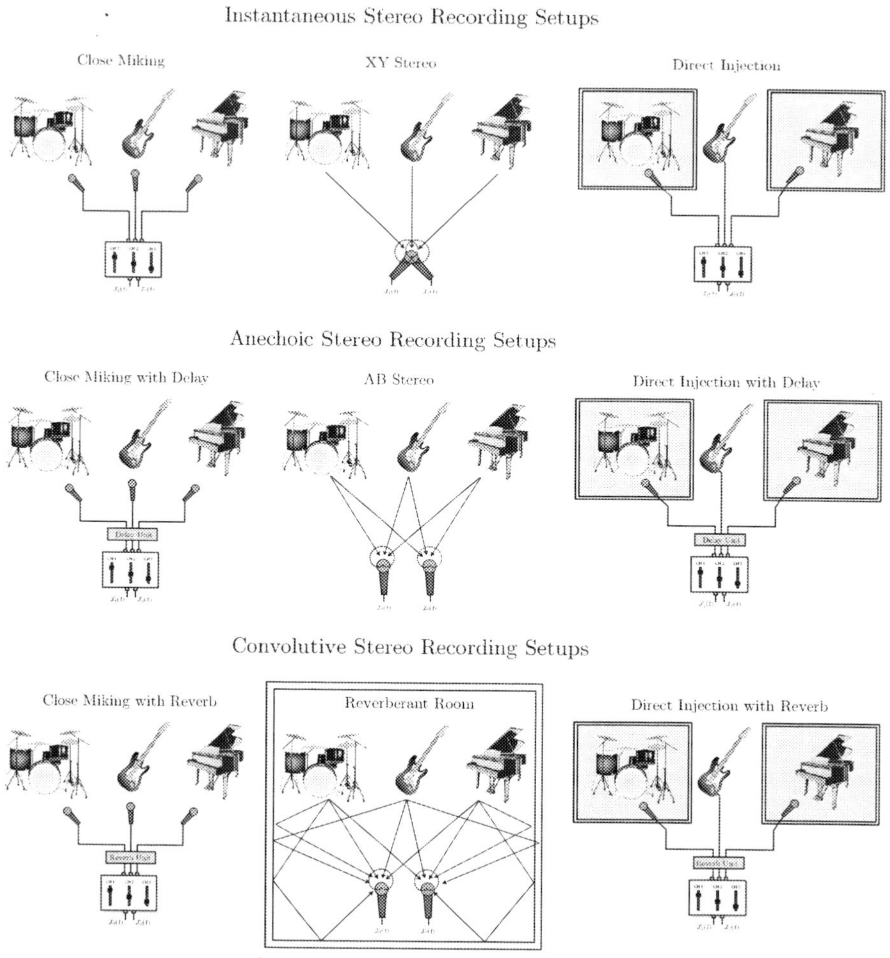

Figure 2. Different mixing set-ups lead to different types of mixtures.

modified by the mixing process. In the following subsections we will describe the different models in source separation and how the images of the sources vary depending on the type of mixing considered.

2.3.1. Instantaneous Model

The simplest mixing model is the *instantaneous model*. In this model, the mixtures are formed by linear combinations of the sources. Therefore, the mixtures are obtained by summing scaled versions of the sources:

$$x_m(t) = \sum_{n=1}^{N} a_{mn} s_n(t), \quad m = 1, \ldots, M, \tag{2}$$

where $x_m(t)$ is the signal of the m-th channel at time sample t, $s_n(t)$ is the n-th source signal and a_{mn} are scalars. The images of the sources are then given by

$$s_{mn}(t) = a_{mn} s_n(t). \tag{3}$$

2.3.2. Anechoic Model

A model that considers only the direct path from each source to each sensor is the *anechoic model*. This model can be thought as an extension of instantaneous mixing in which transmission delays between sensors are also considered. The generative model is

$$x_m(t) = \sum_{n=1}^{N} a_{mn} s_n(t - \delta_{mn}), \quad m = 1, \ldots, M, \tag{4}$$

where δ_{mn} is the arrival delay between source s_n and sensor m and a_{mn} is the attenuation factor corresponding to the path between source s_n and microphone m. The images of the sources are scaled and delayed versions of the sources:

$$s_{mn}(t) = a_{mn} s_n(t - \delta_{mn}). \tag{5}$$

2.3.3. Convolutive Model

In the convolutive or echoic case, reverberation is considered too. Mathematically, the process can be thought as a finite impulse response (FIR) *convolutive model*, written as:

$$x_m(t) = \sum_{n=1}^{N} \sum_{\tau=0}^{L_{imp}-1} h_{mn}(\tau) s_n(t - \tau), \quad m = 1, \ldots, M, \tag{6}$$

where $h_{nm}(\tau)$ is the impulse response of the acoustic path from source n to sensor m, and L_{imp} is the maximum length of all impulse responses. The images of the sources are therefore filtered versions of the original sources:

$$s_{mn}(t) = \sum_{\tau=0}^{L_{imp}-1} h_{mn}(\tau) s_n(t - \tau). \tag{7}$$

2.3.4. Matrix Formulation

The above mixing models are usually described in matrix notation. With this notation, the instantaneous model is mathematically expressed as:

$$\begin{bmatrix} x_1(t) \\ x_2(t) \\ \vdots \\ x_M(t) \end{bmatrix} = \begin{bmatrix} a_{11} & a_{12} & \cdots & a_{1N} \\ a_{21} & a_{22} & \cdots & a_{2N} \\ \vdots & \vdots & \ddots & \vdots \\ a_{M1} & a_{M2} & \cdots & a_{MN} \end{bmatrix} \cdot \begin{bmatrix} s_1(t) \\ s_2(t) \\ \vdots \\ s_N(t) \end{bmatrix}, \tag{8}$$

which is equivalent to the form $\mathbf{x}^t = \mathbf{A}\mathbf{s}^t$. The matrix $\mathbf{A}$ is known as the *mixing matrix*, which has $M \times N$ dimensions. If all time instants are directly considered, the model is expressed as $\mathbf{X} = \mathbf{AS}$, where $\mathbf{X}$ is the $M \times T$ matrix corresponding to the sensor data at all times $t = 1, \cdots, T$ (i.e. row i of $\mathbf{X}$, denoted by $\mathbf{X}_i$, corresponds to the ith mixture signal) and $\mathbf{S}$ is the $N \times T$ matrix of source signals.

When the mixing is anechoic or convolutive, the mixtures are described by $\mathbf{x}^t = \mathbf{A} * \mathbf{s}^t$ and $\mathbf{A}$ can be written as

$$\mathbf{A} = \begin{bmatrix} a_{11}\delta(t-\delta_{11}) & a_{12}\delta(t-\delta_{12}) & \cdots & a_{1N}\delta(t-\delta_{1N}) \\ a_{21}\delta(t-\delta_{21}) & a_{22}\delta(t-\delta_{21}) & \cdots & a_{2N}\delta(t-\delta_{2N}) \\ \vdots & \vdots & \ddots & \vdots \\ a_{M1}\delta(t-\delta_{M1}) & a_{M2}\delta(t-\delta_{M2}) & \cdots & a_{MN}\delta(t-\delta_{MN}) \end{bmatrix}, \quad (9)$$

in the anechoic case and

$$\mathbf{A} = \begin{bmatrix} h_{11}(t) & h_{12}(t) & \cdots & h_{1N}(t) \\ h_{21}(t) & h_{22}(t) & \cdots & h_{2N}(t) \\ \vdots & \vdots & \ddots & \vdots \\ h_{M1}(t) & h_{M2}(t) & \cdots & h_{MN}(t) \end{bmatrix}, \quad (10)$$

when the mixing is convolutive.

2.3.5. Models in the Time-Frequency Domain

Many algorithms make use of the above models in the time-frequency domain, by using the complex Short-Time Fourier Transform (STFT). Time-frequency decompostions are used to determine the spectral content of local sections of a signal as it changes over time. As it has been discussed, audio signals are non-stationary, thus, time-frequency transformations provide a useful approach to audio signal processing. The L-point STFT of a time-domain signal $x_m(t)$ sampled at frequency f_s is given by:

$$X_m(k,l) = \sum_{r=-\frac{L}{2}}^{\frac{L}{2}} x_m(l+r)\mathrm{win}(r)e^{(-j2\pi f_k r)}, \quad (11)$$

where k is a frequency index corresponding to $f_k \in \{0, \frac{1}{L}f_s, \ldots, \frac{L-1}{L}f_s\}$, $\mathrm{win}(r)$ is a window that tapers smoothly to zero at each end, such as a Hanning window, $\frac{1}{2}(1+\cos\frac{2\pi r}{L})$ and l is the new time index.

Given the linearity of the Short-Time Fourier Transform (STFT), the model of equation (6) becomes:

$$X_m(k,l) \approx \sum_{n=1}^{N} H_{mn}(k)S_n(k,l) = \sum_{n=1}^{N} S_{mn}(k,l), \quad m = 1, \cdots, M, \quad (12)$$

where $H_{mn}(k)$ is the frequency response from source n to sensor m, and $S_n(k,l)$ is the STFT of the source $s_n(t)$. The indices k and l denote the frequency index and the time index, respectively. Note that there are two main advantages of working with STFT representations: the first one is that convolutive mixtures can be approximated as instantaneous mixtures at each frequency and the second one is that the sparseness is higher. We will later discuss the role of sparse representations for source separation.

In the case of the simplified anechoic model of equation (4), the STFT representation of the observation channels can be written as

$$X_m(k, ml) = \sum_{n=1}^{N} a_{mn} S_n(k, l) e^{-j2\pi f_k \delta_{mn}}, \quad m = 1, \cdots, M, \tag{13}$$

where $S_n(k, l)$ denotes the STFT of the source s_n, f_k is the frequency corresponding to frequency index k and δ_{mn} is the time delay of the path from source n to sensor m.

The STFT instantaneous model from Eq.(2) is:

$$X_m(k, l) = \sum_{n=1}^{N} a_{mn} S_n(k, l), \quad m = 1, \ldots, M. \tag{14}$$

2.4. Types of Mixtures

The separation difficulty is mainly related to three different aspects: the relative number of mixture channels and sources, the length of the mixing filters and the time variation of the mixing filters [9]. These three criteria are used to characterize the mixtures:

- Relative number of mixture channels and sources:
    1. $M > N$ *Over-determined mixture.*
    2. $M = N$ *Determined mixture.*
    3. $M < N$ *Under-determined mixture.*
- Mixing filters:
    1. Scalars (zero delay): *Instantaneous mixture.*
    2. Scalars and/or Delays (possibly fractional): *Anechoic mixture.*
    3. Otherwise: *Convolutive mixture.*
- Time variation of the mixing filters:
    1. Static sources or fixed filters: *Time-invariant mixture.*
    2. Moving sources or time-varying filters: *Time-variant mixture.*

Note that the formulation of the mixing processes described in Section 2.3. only considers static sources, i.e. fixed filters. Musical mixtures are usually underdetermined, since the most popular format for music recordings is stereo. Therefore, the number of sources is usually higher than the number of channels. Overdetermined and determined situations usually appear in microphone array processing techniques, which are usually used for localization and tracking of the sources [20].

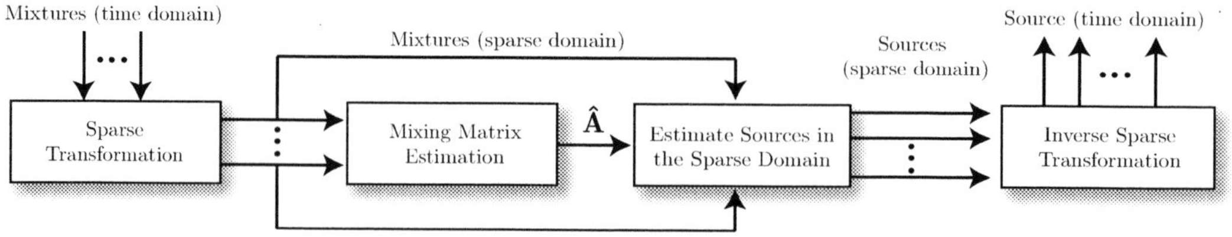

Figure 3. Staged approach for underdetermined source separation.

2.5. Separation Tasks

The SSS problem consists of estimating the source spatial images of the sources $s_{mn}(t)$, from the mixture signals $x_m(t)$. Note that the estimation of the single-channel source signals $s_n(t)$ involves undoing the filtering effect of the mixing process (dereverberating), which is an additional problem that will not be considered in the following.

The instantaneous mixing model, $\mathbf{X} = \mathbf{AS}$, has the form of a conventional system of linear equations. Therefore, it seems that the problem of extracting the sources $\mathbf{S}$ from the mixtures $\mathbf{X}$ can be completely solved by traditional algebraic techniques. This is only possible if the mixing matrix $\mathbf{A}$ is known. However, source separation tries to give solution to this problem in the case were both $\mathbf{S}$ and $\mathbf{A}$ are unknown. Moreover, even if the mixing matrix $\mathbf{A}$ can be accurately estimated, the system is only invertible if the number of equations equals the number of unknowns ($M = N$), i.e. $\mathbf{A}$ is square. When the problem is overcomplete ($M > N$), dimensionality reduction techniques are usually employed. If the problem is underdetermined ($M < N$), there is an infinite number of solutions and demixing the real sources from the mixtures becomes a very difficult task.

Most separation methods follow a two-stage approach, where first estimate the mixing matrix and then estimate the sources. When $\mathbf{A}$ is square (determined problem), the sources can be directly obtained by

$$\hat{\mathbf{S}} = \hat{\mathbf{A}}^{-1}\mathbf{X} = \hat{\mathbf{W}}\mathbf{X}, \tag{15}$$

and the source separation problem is reduced to the estimation of the mixing matrix and its inversion ($\mathbf{W}$).

Independent Component Analysis (ICA) addresses the problem by assuming that the sources are statistically independent and have a non-gaussian distribution. The sources can be obtained up to a permutation and scaling ambiguity (the energies and order of the sources cannot be determined). A key concept in ICA algorithms is nongaussianity. The Central Limit Theorem states that the distribution of a sum of independent random variables tends toward a gaussian distribution, under certain conditions. Thus, a sum of two independent random variables usually has a distribution that is closer to gaussian than any of the original random variables. Therefore, the maximization of a nongaussianity measure leads to the independent components. Classical quantitative measures of nongaussianity are kurtosis and negentropy. Alternative approaches to ICA are those based on minimization of mutual information and maximum-likelihood estimation. For a good review on Independent Component Analysis see [21].

In the underdetermined case, $\mathbf{A}$ is rectangular and thus non-invertible, which forces to perform a second stage after the mixing matrix estimation stage. Most approaches for underdetermined source separation are based on sparse transformations of the mixtures. In the next section, we review some of the ideas underlying these sparse methods for the separation of two-channel mixtures.

3. Stereo Separation

3.1. Sparsity

A signal is said to be sparse if most of its components are zero or close to zero. Sparsity is desirable for source separation because, the sparser a signal is, the less it will overlap with the other signals in a mixture [22]. Consider, for example, the following instantaneous mixture with only two sources:

$$\begin{bmatrix} x_1(t) \\ x_2(t) \end{bmatrix} = \begin{bmatrix} a_{11} & a_{12} \\ a_{21} & a_{22} \end{bmatrix} \cdot \begin{bmatrix} s_1(t) \\ s_2(t) \end{bmatrix}. \tag{16}$$

If only one of the sources is active, for example s_1, the resultant mixture will be

$$\begin{bmatrix} x_1(t) \\ x_2(t) \end{bmatrix} = s_1(t) \begin{bmatrix} a_{11} \\ a_{21} \end{bmatrix}, \tag{17}$$

and so, the points of the scatter plot of $x_1(t)$ versus $x_2(t)$ would lay on a line crossing the origin with direction given by the vector $[a_{11}\ a_{21}]^T$, showing a very clear structure imposed by the linear mixing. Therefore, if the sources are active in few points, i.e. they are sparse, there will be low probability for more than one source being active at the same time and the scatter plot would constitute a mixture of lines following the directions given by the columns of $\mathbf{A}$, denoted by $\mathbf{a}_j$. Sparse methods estimate these line orientations (also known as basis vector) from the observed data. Thus, the sparser the signals, the more their coefficients will be concentrated around the mixing directions and the easier will be the detection of line orientations. Figure 4 shows the scatter plot of an instantaneous stereo mixture of four sources in the time domain and in the STFT domain. It can be clearly observed that points are clustered along four line directions in the time-frequency domain, which are directly related to the mixing ratios of the sources. Sparse methods are very different from traditional ICA approaches: while ICA exploits the statistics of the sources, sparse methods make use of the structure of the mixtures. Once the mixing parameters of the sources have been estimated, the sources are usually demixed by applying clustering or optimization techniques.

3.2. Sparse Transformations

Under a sparse signal representation, the probability of two or more sources being simultaneously active is low, making easier the separation task. The *disjointness* of a mixture can be defined as the degree of non-overlapping of the mixed signals. Generally, a higher sparsity will result in a higher disjointness and in an easier separation. As has been depicted in Figure 4, most audio mixtures are not sufficiently disjoint in the time domain and it is often required to transform the signals into a sparser domain, such as a time-frequency

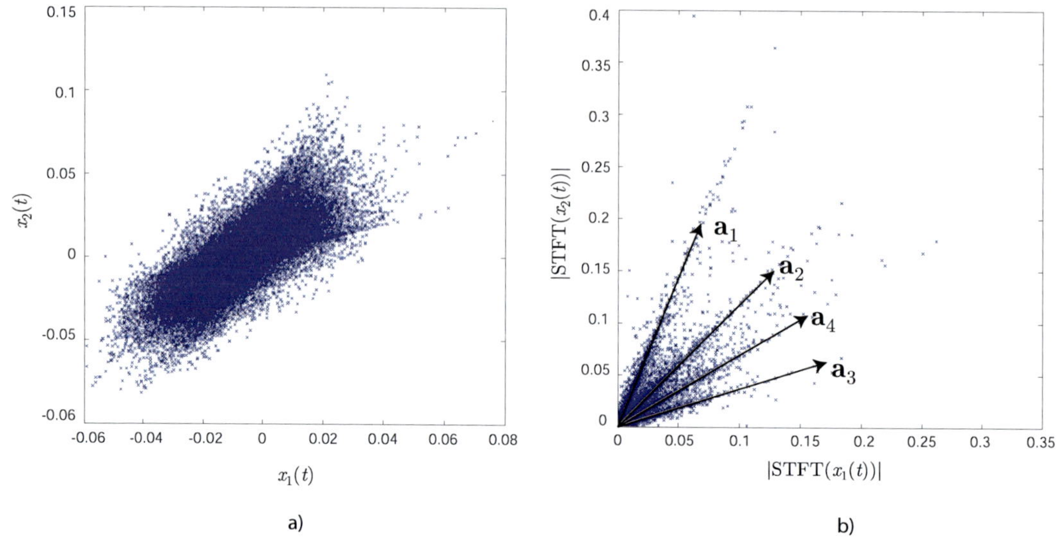

Figure 4. Scatter Plot of an instantaneous stereo mixture with 4 sources. a) Time Domain. b) STFT domain.

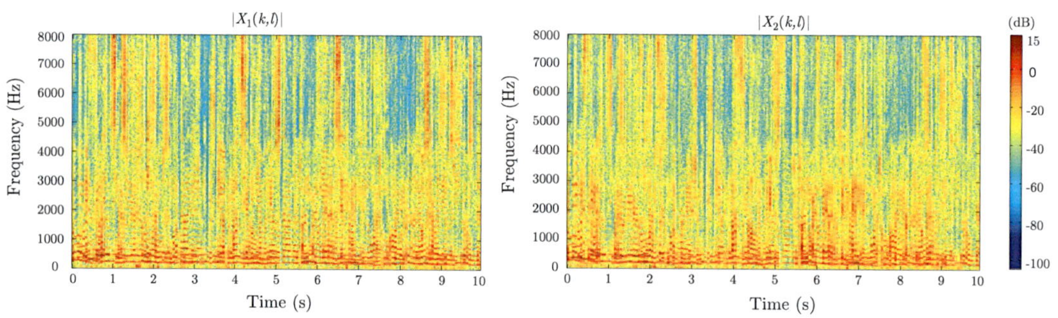

Figure 5. STFT magnitude of an instantaneous stereo mixture with 4 speech sources.

representation. Wavelets [23], overcomplete dictionaries [24], atomic decompositions [25], auditory representations [22] or the Modified Discrete Cosine Transform (MDCT) [26] have been proposed in the literature as alternatives for obtaining sparsity. One of the most utilized transformations in underdetermined source separation methods is the STFT. It has been shown that the STFT provides higher disjointness for music signals than for speech signals. This can be explained by the fact that speech concentrates its energy in a narrower part of the spectrum, and therefore spectral overlapping is more likely to occur. In Figure 5, the STFT magnitude of the mixtures of four speakers analyzed in Figure 4 are represented. As can be seen, most of the energy is concentrated in harmonic partials in the mid-frequency range and many TF points have values below -20 dB.

Generally, it can be assumed that audio signals have a sparse representation in a proper signal dictionary φ:

$$s_n(t) = \sum_{u=1}^{U} C_{nu}\varphi_u(t), \tag{18}$$

where the functions $\varphi_u(t)$ are the *atoms* or *elements* of the dictionary and C_{nu} are weighting coefficients. The elements φ_u do not have to be linearly independent and they may form an overcomplete dictionary. Therefore, under this notation, sparsity means that only a small number of the coefficients C_{nu} are significantly different from zero. If a whole dictionary matrix Φ with dimensions $U \times T$ is considered, the matrix of sources $\mathbf{S}$ can be expressed as

$$\mathbf{S} = \mathbf{C}\Phi, \tag{19}$$

where $\mathbf{C}$ is a $N \times U$ matrix of coefficients.

The *Maximum A Posteriori* (MAP) approach for source separation leads to several sparsity-based methods. It consists in maximizing the posterior distribution $P(\mathbf{A}, \mathbf{C}|\mathbf{X})$:

$$\max_{\mathbf{A},\mathbf{C}} P(\mathbf{A}, \mathbf{C}|\mathbf{X}) \propto \max_{\mathbf{A},\mathbf{C}} P(\mathbf{X}|\mathbf{A}, \mathbf{C})P(\mathbf{A})P(\mathbf{C}). \tag{20}$$

A number of probability density functions can be considered for $\mathbf{C}$, being the laplacian distribution one of the most utilized priors. If the mixing matrix is firstly estimated following a staged approach, the MAP problem becomes a l_1 norm minimization problem:

$$\hat{\mathbf{S}} = \min_{\mathbf{S}} \left(\sum_{n,t} |s_n(t)| \right) \quad \text{subject to} \quad \mathbf{X} = \hat{\mathbf{A}}\mathbf{S}. \tag{21}$$

A solution for the above problem in the case of $M = 2$ observation channels was proposed by Bofill and Zibulevski [27], known as the *shortest-path algorithm.*

3.3. Time-Frequency Masking

Time-frequency masking is a powerful approach for the separation of underdetermined mixtures. Techniques based on time-frequency masking use a time-frequency representation of the signal, taking profit from the disjointness provided by sparse transformations, such as the STFT or a cochleagram, which is later described in Section 4.2.1.. Their aim is to identify the dominating source in each time-frequency unit, obtaining a mask that indicates which are the active points of each source in the time-frequency domain. Most common techniques are usually based either on a CASA approach (using perceptual principles of auditory scene analysis) or on blind separation methods.

From the CASA perspective, a cochleagram of a mixture is partitioned on contiguous time-frequency regions (called segments), which are later grouped into streams corresponding to different sound sources. The result of the grouping is usually a binary mask that indicates which parts of the input belong to the source to be separated. The advantage of this approach is that separation can be performed using only one observation channel. However, these methods are aimed at separating only one target source from a mixture.

Blind separation methods based on time-frequency masking are also popular. Yilmaz and Rickard [28] applied binary time-frequency masks to mixtures of several speech sources in the STFT domain considering a two-sensor arrangement. They observed that speech sources are sufficiently disjoint under time-frequency representations and therefore, their algorithm is based on estimating which time-frequency points belong to each of the sources in the mixture. In addition, their algorithm considers relative delays between the sources, which allows it to separate audio sources in anechoic scenarios. This algorithm, known as DUET, assumes that the sources are *W-disjoint Orthogonal* (W-DO). Under this assumption, it is likely that every time-frequency point in the mixture with significant energy is dominated by the contribution of one source. This can be mathematically expressed as

$$S_x(k,m)S_y(k,m) = 0, \forall y \neq x, \forall k, m, \tag{22}$$

where $S_x(k,m)$ and $S_y(k,m)$ are the STFT of two any sources in the mixture.

The set of masks estimated by a time-frequency masking algorithm can be applied to the mixtures in order to obtain the estimates of the sources:

$$Y_{ij}(k,m) = \hat{S}_{ij}(k,m) = M_j(k,m)X_i(k,m), \tag{23}$$

being $Y_{i,j}(k,m)$ the STFT of the image of source s_j in sensor i and $M_j(k,m)$ the separation mask for that source. The estimates of the sources in the time domain are obtained applying the inverse STFT operator.

In practice, perfect separation is very difficult to be achieved and the estimated source spatial images may contain different distortions: musical noise, interference from other sources, timbre distortion and spatial distortion. Musical noise, which consists of burbling artifacts appearing with time-frequency masking algorithms, is one of the most common distortions in source separation.

3.3.1. The DUET Algorithm

One of the most popular algorithms for stereo underdetermined mixtures is the DUET algorithm [29]. Each observation channel is transformed into the STFT time-frequency domain, and the relative attenuation and delay values between two observations are calculated from the ratio of the corresponding time-frequency points R_{21}, written as

$$R_{21}(k,l) = \frac{X_2(k,l)}{X_1(k,l)}. \tag{24}$$

The symmetric attenuation estimate is defined as

$$\hat{\alpha}(k,l) = |R_{21}(k,l)| - \frac{1}{|R_{21}(k,l)|} = \hat{a}(k,l) - \frac{1}{\hat{a}(k,l)}, \tag{25}$$

and the instantaneous delay estimate is defined as

$$\hat{\delta}(k,l) = -\frac{1}{\omega_k}\angle R_{21}(k,l), \tag{26}$$

where ω_k is the angular frequency associated to frequency index k. The relative mixing parameters, $\hat{a}_n$ and $\hat{\delta}_n$, for each source in the mixture are estimated from peaks in a 2D power weighted histogram constructed from $\hat{\alpha}(k,l)$ and $\hat{\delta}(k,l)$. Figure 6 shows the histogram obtained for the example mixture, where the 4 sources can be clearly observed as prominent peaks that give an estimation of the mixing parameters. A set of separation masks $M_n(k,l)$ is obtained assigning each time-frequency point to the closest mixing parameter estimate.

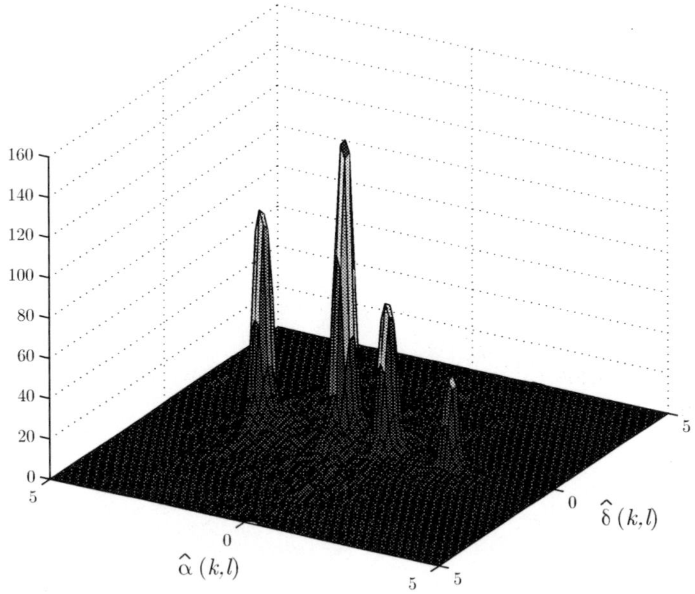

Figure 6. 2D Power weighted histogram for estimating the separation masks. As the mixture is instantaneous, all the peaks are aligned in the value $\hat{\delta} = 0$.

The separation of the sources is carried out using time-frequency masking and a Maximum Likelihood criterion:

$$\hat{S}_n(k,l) = M_n(k,l)\left(\frac{X_1(k,l) + \hat{a}_n e^{j\hat{\delta}_n \omega} X_2(k,l)}{1 + \hat{a}_n^2}\right). \tag{27}$$

The resultant masks for each of the sources are depicted in Figure 7, where white indicates a value of 1 and black indicates a value of 0. It is important to remark that a phase ambiguity problem may appear in the upper frequency range: the inter-microphone spacing d must satisfy the condition $d < c/f_s$ to avoid the spatial aliasing problem, being f_s the sampling frequency. Therefore, it is guaranteed that there is less than one wavelength difference between two sources with a direction-of-arrival of $+\pi/2$ and $-\pi/2$ from the equal-delay direction.

4. Monaural Separation

Blind Audio Source Separation methods are generally based on statistical assumptions or spatial structure of the sources. The use of several observation channels make possible to estimate several sources with very little information about them. However, the use of advanced source models can improve the separation quality or even perform separation from monaural mixtures. Therefore, some a priori knowledge is necessary to extract the sources from a single mixture. Musical instruments have a wide range of sound production mechanisms, and the resultant signals have a wide range of spectro-temporal characteristics. Moreover, the statistical assumption regarding the independency of the sources is not always completely fulfilled, as even though the acoustic signals are produced independently in each

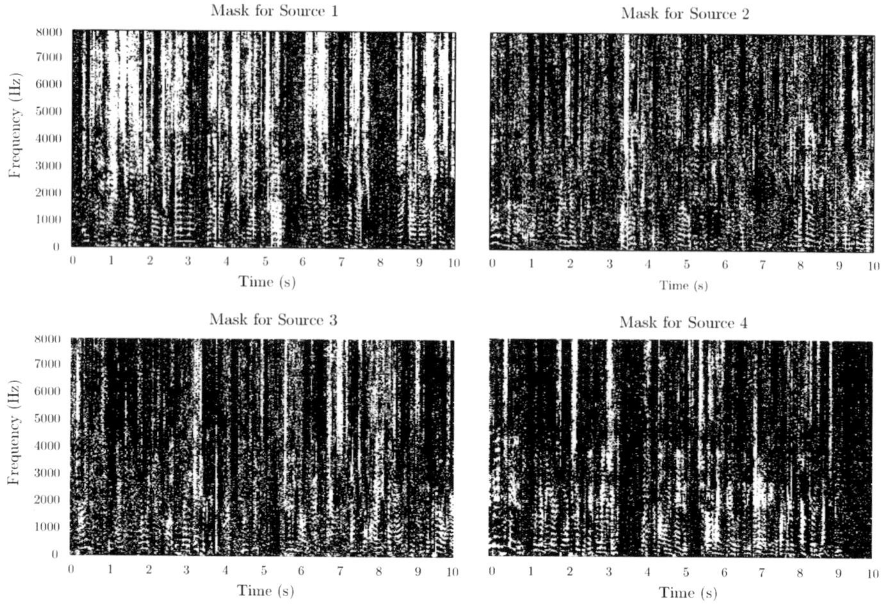

Figure 7. DUET separation masks.

source, it is their interplay which makes up the music. Algorithms that belong to these one-channel methods are often classified between *supervised* and *unsupervised* approaches [30]:

- *Supervised*: Include methods based on training the model with a sound example database.

- *Unsupervised*: No source-specific prior knowledge.

Unsupervised methods are also divided into three categories:

- *Model-based Inference*: This methods use a parametric model of the sources and the parameters are estimated from the observed mixture signal. Implicit information can be used to design different types of estimators. In music applications, the most commonly used parametric model is the sinusoidal model, suitable for the separation of pitched musical instruments.

- *Unsupervised Learning*: These methods usually apply a simple non-parametric model and use less prior information of the sources. They learn the source characteristics from the data. Some algorithms are based on the independence and sparsity assumptions already discussed. One of the most popular approaches here is *Non-Negative Matrix Factorization* (NMF).

- *Psychoacoustically motivated methods*: This category includes almost all CASA methods based on auditory perception.

4.1. Non-negative Matrix Factorization

Consider a single signal mixture of several sources. Let us express the observed mixture in each time frame $\mathbf{x}^t$ as a weighted sum of basis functions $\mathbf{b}_j$, $j = 1 \dots J$, so that:

$$\mathbf{x}^t = \sum_{j=1}^{J} g_{j,t}\mathbf{b}_j \quad t = 1, \dots, T, \tag{28}$$

where J is the number of basis functions, and $g_{j,t}$ is the gain of the j-th basis function in the t-th frame. Following this model, each sound source can be modeled as a sum of several components (where the term component denotes here a basis function together with its time varying gain):

$$\mathbf{y}_n^t = \sum_{j \in \mathcal{S}_n} g_{j,t}\mathbf{b}_j, \tag{29}$$

where $\mathcal{S}_n$ is the set of components within source n. In a matrix form, the model becomes:

$$\mathbf{X} = \mathbf{BG}, \tag{30}$$

where $\mathbf{B} = [\mathbf{b}_1, \mathbf{b}_2, \dots, \mathbf{b}_J]$ is the *basis matrix*, and $\mathbf{G}$ is the *gain matrix*. Although several data representations can be used for the observed signals, absolute values of the DFT are often used. Therefore, $\mathbf{x}^t$ is the magnitude spectrum in frame t, and each component j has a fixed magnitude spectrum $\mathbf{b}_j$ with a time-varying gain $g_{j,t}$. As magnitude spectra are non-negative by definition, it is advantageous to restrict the basis functions to be entry-wise non-negative. In addition, it is useful to allow only positive gains (components purely additive). Thus, in the signal model $\mathbf{X} = \mathbf{BG}$, the element-wise non-negativity of $\mathbf{B}$ and $\mathbf{G}$ alone is sufficient condition for the separation of sources in many cases (without an explicit assumption of their independence).

The *non-negative matrix factorization* (NMF) algorithm is based on the following optimization problem:

$$\min \|\mathbf{X} - \mathbf{BG}\|^2 \quad \text{subject to} \quad \mathbf{B}, \mathbf{G} \geq 0. \tag{31}$$

The algorithm of Lee and Seung [8] minimizes the Euclidean distance by initializing the entries of $\mathbf{B}$ and $\mathbf{G}$ with random positive values, and then by updating them iteratively using multiplicative rules:

$$\mathbf{B} \leftarrow \mathbf{B}. \times (\mathbf{XG}^T)./(\mathbf{BGG}^T) \tag{32}$$

and

$$\mathbf{G} \leftarrow \mathbf{G}. \times (\mathbf{B}^T\mathbf{X})./(\mathbf{B}^T\mathbf{BG}), \tag{33}$$

where $.\times$ and $./$ denote the element-wise multiplication and division, respectively, and ^{T} denotes matrix transposition.

To summarize, the NMF algorithm is composed of the next steps:

1. Initialize each entry of $\mathbf{B}$ and $\mathbf{G}$ with the absolute values of Gaussian noise.
2. Update $\mathbf{G}$ using Eq.(33).
3. Update $\mathbf{B}$ using Eq.(32).

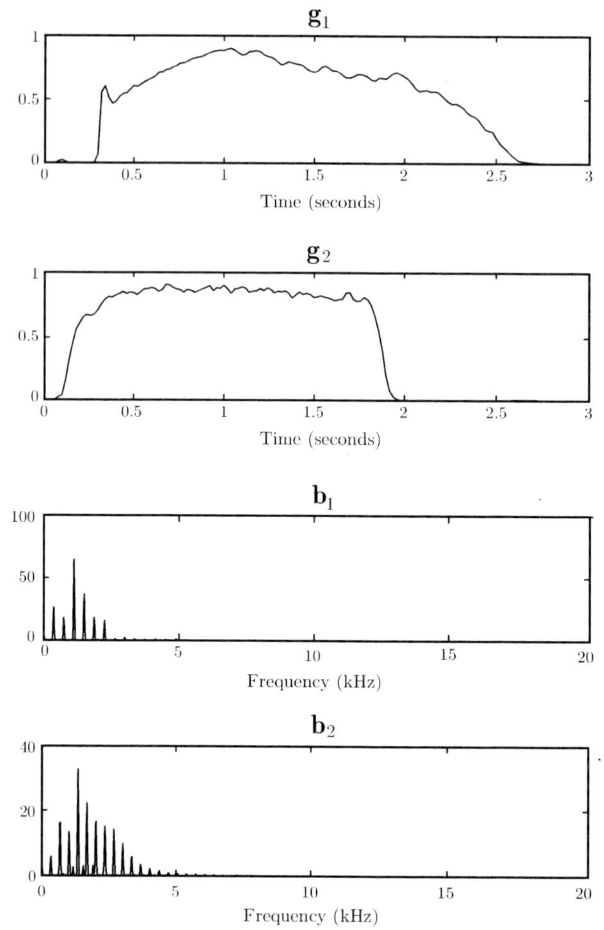

Figure 8. NMF components estimated from a mixture signal of two notes (trumpet and oboe). Gains are plotted on top and basis functions at the bottom.

4. Repeat Steps 2 and 3 until the values converge.

Note that the NMF can be used only for a non-negative observation matrix, thus, it is not suitable for separating time-domain signals. If a mixture spectrogram is a sum of sources with a static spectrum with a time-varying gain, and each of them is active in at least one frame and frequency in which the other components are not active, NMF provides perfect separation. However, real-world music signals rarely fulfill these conditions, and two or more sources that are simultaneous at all times would be represented by a single component. NMF with a sparsness constraint is referred as *non-negative sparse coding* (NSC). This technique uses also minimization of Euclidean distances but adding an additional term that enforces sparseness on the activity of the basis functions. For example, Virtanen [30] proposed NSC for monaural sound separation based on minimizing a cost function $c(\mathbf{B}, \mathbf{G})$, which is a weighted sum of three terms: a reconstruction error term (Euclidean distance) $c_r(\mathbf{B}, \mathbf{G})$, a temporal continuity term $c_t(\mathbf{G})$, and a sparseness term $\beta c_s(\mathbf{G})$:

$$c(\mathbf{B}, \mathbf{G}) = c_r(\mathbf{B}, \mathbf{G}) + \alpha c_t(\mathbf{G}) + \beta c_s(\mathbf{G}), \tag{34}$$

where α and β are the weights for the temporal continuity term and sparseness term, respectively. The cost functions for temporal continuity is:

$$c_t(\mathbf{G}) = \sum_{j=1}^{J} \frac{1}{\sigma_j^2} \sum_{t=2}^{T} (g_{t,j} - g_{t-1,j})^2, \tag{35}$$

where σ_j is the standard deviation of the gains. This cost function penalizes large changes between the gains $g_{j,t}$ and $g_{j,t-1}$ in adjacent frames.

The cost function for sparseness is:

$$c_s(\mathbf{G}) = \sum_{j=1}^{J} \sum_{t=1}^{T} f(g_{j,t}/\sigma_j), \tag{36}$$

where $f()$ is a function which penalizes non-zero gains, for example, $f(x) = |x|$.

Figure 8 shows the result of applying the algorithm to a signal made up of two notes played by two different instruments (trumpet and oboe). Two components were calculated corresponding to the harmonic spectrum of the two notes played. Their temporal gain is showed in the two upper plots and they give information about when the notes are being played and its level over time.

4.2. Computational Auditory Scene Analysis

In this subsection, we introduce some of the fundamental ideas associated with CASA systems [3]. Computational systems for auditory scene analysis usually follow a similar structure. Firstly, the recorded mixture enters the peripheral analysis block, which results in a time-frequency representation of auditory activity. The most common front-end used in CASA systems is the cochleagram. Using this representation, an analysis step is carried out with the aim of extracting relevant features, such as periodicity, onsets, offsets, frequency and amplitude modulation, etc. With the aid of these features, mid-level representations are formed (segments). The grouping cues of Auditory Scene Analysis combined with specific source models are then used to perform separation of the mixture into different streams. In the last block, the final audio waveform is resynthesized from the separated streams.

4.2.1. The Cochleagram

The cochleagram is a computer model of the peripheral auditory system that models its human frequency selectivity. This is usual achieved by means of a filterbank, in which each channel models the frequency response associated with a particular point of the basilar membrane. In this context, the gammatone filter is a popular model for the impulse response function of auditory nerve fibers. It is a bandpass filter, whose frequency response is:

$$G(f) \approx \left[1 + \frac{j(f - f_c)}{b(f_c)}\right]^{-N} \quad 0 < f < \text{inf}, \tag{37}$$

where N is the filter order, f_c is the center frequency (in Hz) and $b(f_c)$ is a function that determines the bandwidth for a given center frequency, which is usually set according to

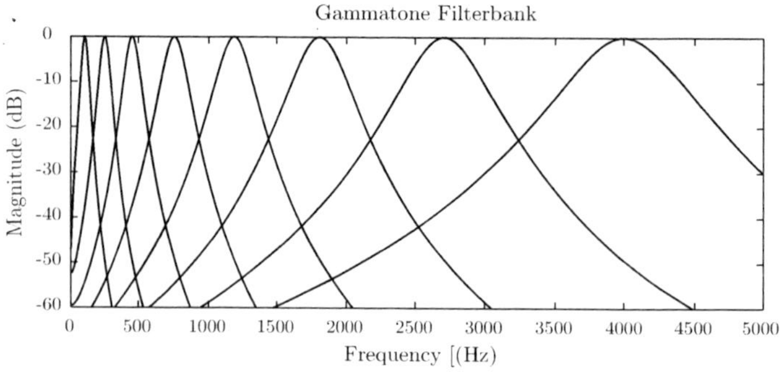

Figure 9. Amplitude response of a gammatone filterbank.

measurements of the equivalent rectangular bandwidth (ERB) of human auditory filters $b(f) = 1.019\,\mathrm{ERB}(f)$, with

$$\mathrm{ERB}(f) = 24.7 + 0.108f. \tag{38}$$

The response of a gammatone filterbank is represented in Figure 9. The output of the filter may be followed by a block aimed at simulating the auditory nerve activity, such as de Meddis model [31]. Thus, a complete cochlear model includes a gammatone filterbank and a Meddis hair cell model, resulting in a time frequency representation termed as cochleagram.

4.2.2. Correlogram

The correlogram can be interpreted as a model of pitch perception. It is computed as:

$$acf(t, c, \tau) = \sum_{k=0}^{K-1} a(t-k, c)a(t-k-\tau, c)h(k), \tag{39}$$

where $a(t, c)$ is a simulated auditory nerve response for channel c at discrete time t, and τ is a time lag. The autocorrelation is computed over a window function $h(k)$ of length K samples (usually a Hanning window).

When the input is a pitched sound (a vowel, for example), the lowest frequency channels of the correlogram exhibit peaks at the pitch period of the input, but also at multiples of this period, since the autocorrelation of a periodic signal is itself a periodic function. Those filters driven by higher harmonics also exhibit peaks at multiples of the harmonic period, and obviously, some of them coincide with the peak of the pitch. At high frequencies, auditory filters are wider and a number of harmonics interact within the same filter, resulting in amplitude modulation at the rate of the pitch frequency. Hence, the correlogram of a periodic shows a higher density of peaks at those lags corresponding to the pitch period and its multiples.

The pitch of the signal can be emphasized by computing the *summary correlogram*, given by

$$sacf(t, \tau) = \sum_{c} acf(t, c, \tau). \tag{40}$$

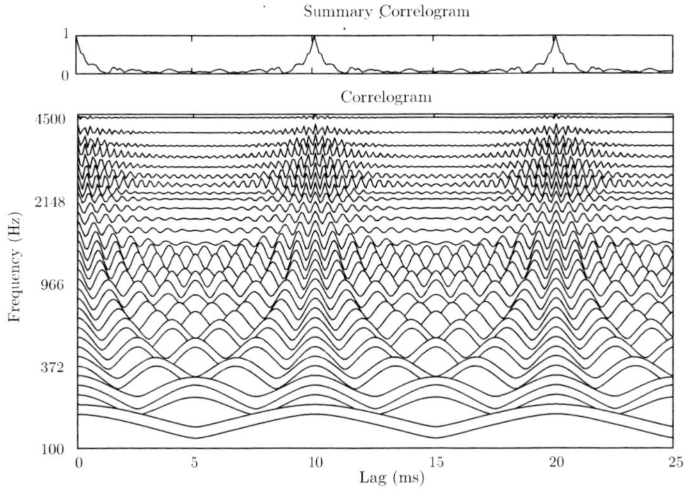

Figure 10. Bottom: Correlogram and summary correlogram for a pitched sound. Top: Summary correlogram where peaks corresponding to multiples of the pitch period can be observed. Bottom: Correlogram.

The summary correlogram exhibits strong peaks at the lag corresponding to the pitch period and its multiples. Therefore, the peak that occurs at the shortest lag corresponds to the fundamental period. An example correlogram and summary correlogram are depicted in Figure 10.

4.2.3. Feature Extraction and Grouping

Auditory Scene Analysis studies have shown that every listener makes use of some common properties of the auditory system. These features include proximity in frequency and time, temporal continuity, periodicity, onset and offset, frequency modulation and spatial location. After computing some of these features, CASA methods obtain a set of segments that form contiguous regions on the cochleagram regarding the extracted features. A segment is often defined as an intermediate representation that its underlying acoustic energy primarily originates from the same sound source. For example, a segment may correspond to a resolved harmonic of a speech source, a formant or a fricative sound. After segmentation, a simultaneous grouping step is carried out for grouping segments overlapped in time, which results in a set of streams. The streams originated are later grouped in a sequential grouping step with the aim of linking those originated from the same sound source. For example, grouping successive speech sections of the same speaker. This sequential grouping is usually performed via an spectrum-based approach, assuming that spectral features of the same speaker are more similar than those of different speakers.

4.2.4. Binary Masking and Resynthesis

The time-frequency masking approach of Section 3.3. is one of the commonest techniques used in CASA. In fact, the ideal binary mask (IBM) has been suggested as a computational goal for these methods. This ideal mask has non-zero values at time-frequency

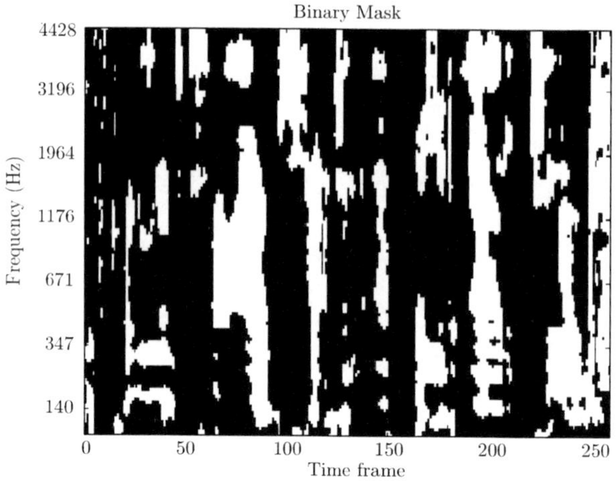

Figure 11. Ideal binary mask for extracting a speech source from a monaural mixture.

points where the energy of the target source is greater than the interfering energy. A typical CASA binary mask is shown in Figure 11.

The resynthesis step allows to reconstruct an audio waveform from the grouped streams. Resynthesis is usually achieved by inverting the time-frequency transformation. First, across-channel phase shifts in the filterbank output are removed by reversing the response of each filter in time, passing the reversed response back through the filter, and then time-reversing the filtered response again. Alternatively, the phase-corrected output from each filter channel is then divided into time frames by windowing with a frame size that matches the original decomposition. The energy in each time-frequency unit is weighted by the corresponding mask value. Finally, the weighted responses are then summed across all frequency channels.

5. Applications

5.1. Music Up-Mixing

Over the last few decades, surround sound, or multichannel sound reproduction systems, have played an increasingly important role in the entertainment industry, as well as in the field of multimedia. While high-fidelity audio devices currently demonstrate flat frequency responses and very low levels of noise and distortion, reproduction issues involving spatial quality or spatial fidelity are still an active research field. In fact, multichannel audio systems try to reproduce sound in a way which is more "natural" with the aim of enhancing the listening experience. This is possible due to the improved spatial attributes of these systems, like the ability to facilitate the perception of sound sources coming from different directions, for example.

Despite all of the advances made in spatial sound reproduction over the last few years, the vast majority of musical recordings are stored and supplied in a two-channel "stereo" format, making it necessary to listen to them on a two-loudspeaker reproduction system. In this context, audio signal processing systems for converting stereo recordings into four or five channels are gaining attention. These up-mixers are used for reproducing conventional stereo recordings with spatial reproduction systems, taking advantage of the spatial properties of multichannel audio reproduction. Stereo-to-5.1 up-mixers are usually based on a matrix scheme, which generates the additional channels by simple adding and subtracting the input channels with altered gain and phase. However, source separation techniques can be a solution for the isolation of different sound objects in conventional recordings, allowing to resynthesize acoustic scenes in multichannel sound systems with relatively good quality [32]. Therefore, although separation algorithms are far from giving high fidelity audio signals, the masking effects involved in the listening of spatially remixed scenes perceptually improve the perception of the virtual sources.

5.2. Music Re-processing

Important applications of BASS methods are within music re-processing tools. Here, we understand as re-processing tools as those that are aimed at modifying a musical mixture selectively on auditory streams. For example:

- modifying the volume of a given instrument within a mixture,
- karaoke applications,
- adding specific effects to a given instrument or singer,
- obtaining instrument samples from the original mix with re-mixing purposes.

An interesting application based on a time-frequency masking approach is the plug-in developed by the Music Technology Group of the Pompeu Fabra University, which enables the user to select musical objects with different panning in the stereo mixtures [33].

5.3. Automatic Music Transcription

Musical transcription of audio data is the process of taking a sequence of digital data corresponding to the sound waveform and extracting from it the symbolic information related to the high-level musical structures that might be seen on a score. In this application the sound sources are one or more instruments playing a piece of music. An analysis is carried out to identify the instruments that are playing and to estimate when and for how long each note is played. From this analysis it is possible to produce a written musical score, showing notes and durations in the form of conventional music notation. In principle, this musical score can then be used to recreate the musical piece that was played. Most pieces of music are polyphonic, having one or more instruments playing several notes at once, such as a piece for piano or full orchestra. The use of the term monophonic here should be distinguished from a mono (as opposed to stereo) recording, which may be polyphonic music but has been recorded using a single audio channel. Even the best human music transcribers are typically limited to pieces with less than 5 notes at once. For general listeners, the identity of individual notes does not seem to be important. Thus, the automatic music transcription problem is somewhat artificial: it is not something that we would expect an average human to be able to do easily. However, computational models including source separation may give in the future accurate results under a wide range of musical recordings, resulting in very interesting musical applications.

6. Conclusion

Throughout this chapter, some basic ideas underlying source separation methods have been presented. As music mixtures are usually underdetermined, i.e. there are more sources than observation channels, special attention has been given to sparse methods for stereo separation and to monaural approaches. Generally, it has been shown that there not source separation algorithms that can be successfully applied to any kind of mixture, as they are very dependent on the nature of the recording (number of sources, number of sensors and mixing process). Moreover, these algorithms are still far from providing high-quality separated tracks, hence there is still much work to do towards real audio-quality-oriented applications. Time-frequency masking algorithms can provide separation in real situations without very high computational cost, but the extracted signals are likely to have musical noise and residuals from other sources. Monaural approaches are very interesting because they are able to extract very relevant information without making use of spatial cues, but their application is rather limited, as they are designed to deal with specific types of sources and mixtures. Therefore, it seems that significance-oriented applications, where only some information about the sources is needed but not their original waveform, can take better profit of current source separation methods.

References

[1] E. C. Cherry, "Some experiments on the recognition of speech, with one and with two ears," *Journal of the Acoustical Society of America*, vol. 25, no. 5, pp. 975–979, 1953.

[2] D. P. W. Ellis, "Hard problems in computational auditory scene analysis," Posted to the AUDITORY email list, August 1995.

[3] D. L. Wang and G. J. Brown, Eds., *Computational Auditory Scene Analysis: Principles, Algorithms, and Applications*. Wiley, 2006.

[4] A. S. Bregman, *Auditory Scene Analysis*. MIT Press: Cambridge, MA, 1990.

[5] E. Scheirer, "Music-listening systems," Ph.D. dissertation, Massachusetts Institute of Technology, MA, USA, April 2000.

[6] J. F. Cardoso, "Blind signal separation: Statistical principles," in *Proccedings of the IEEE*, vol. 86, no. 10, October 1998, pp. 2009–2025.

[7] E. Vincent, "Musical source separation using time-frequency source priors," *IEEE Transactions on Audio, Speech and Language Processing*, vol. 14, no. 1, pp. 91–98, 2006.

[8] D. D. Lee and H. S. Seung, "Learning the parts of objects by non-negative matrix factorization," *Nature*, vol. 401, no. 6755, p. 788791, 1999.

[9] P. O'Grady, B. Pearlmutter, and S. Rickard, "Survey of sparse and non-sparse methods in source separation," *International Journal of Imaging Systems and Technology (IJIST)*, vol. 15, no. 1, pp. 18–33, 2005.

[10] E. Vincent, C. Fevotte, R. Gribonval, X. Roder, E. LeCarpentier, L. Benaroya, A. Rbel, and F. Bimbot, "A tentative typology of audio source separation tasks," in *4th International Symposium on Independent Component Analysis and Blind Signal Separation (ICA2003)*, Nara, Japan, April 2003.

[11] M. M. Boone, E. N. G. Verheijen, and P. F. van Tol, "Spatial sound field reproduction by wave field synthesis," *Journal of the Audio Engineering Society*, vol. 43, no. 12, pp. 1003–1012, 1995.

[12] M. A. Gerzon, "With-height sound reproduction," *Journal of the Audio Engineering Society*, vol. 21, pp. 2–10, 1973.

[13] M. Ryynanen, T. Virtanen, J. Paulus, and A. Klapuri, "Accompaniment separation and karaoke application based on automatic melody transcription," in *Proceedings of the IEEE International Conference on Multimedia and Expo*, Hannover, Germany, June 2008, pp. 1417–1420.

[14] S. Y. Low and S. Nordholm, "A hybrid speech enhancement system employing blind source separation and adaptive noise cancellation," in *Proceedings of the 6th Nordic Signal Processing Symposium NORSIG 2004*, Espoo, Finland, June 2004, pp. 204–207.

[15] N. Orio, "Music retrieval: A tutorial and review," *Foundations and Trends in Information Retrieval*, vol. 1, no. 1, pp. 1–90, 2006.

[16] M. D. Plumbley, S. A. Abdallah, J. P. Bello, M. E. Davies, G. Monti, and M. B. Sandler, "Automatic music transcription and audio source separation," *Cybernectics and Systems*, vol. 33, no. 6, pp. 603–627, 2002.

[17] L. Benaroya, F. Bimbot, G. Gravier, and R. Gribonval, "Experiments in audio source separation with one sensor for robust speech recognition," *Speech Communication*, vol. 48, no. 7, pp. 848–854, 2006.

[18] E. Vincent and M. D. Plumbley, "Low bitrate object coding of musical audio using bayesian harmonic models," *IEEE Transactions on Audio, Speech and Language Processing*, vol. 15, no. 4, pp. 1273–1282, 2007.

[19] E. Vincent, M. Jafari, S. Abdallah, M. Plumbey, and M. Davies, "Model-based audio source separation," Queen Mary, University of London, Tech Report C4DM-TR-05-01, November 2005.

[20] J. H. DiBiase, H. F. Silverman, and M. S. Brandstein, *Microphone Arrays: Signal Processing Techniques and Applications*. Springer-Verlag, 2001, ch. Robust Localization in Reverberant Rooms, pp. 157–180.

[21] A. Hyvärinen, J. Karhunen, and E. Oja, *Independent Component Analysis.* John Wiley & Sons, 2001.

[22] J. J. Burred and T. Sikora, "On the use of auditory representations for sparsity-based sound source separation," in *Proceedings of the 5th International Conference on Information, Communications and Signal Processing (ICICS 2005)*, Bangkok, Thailand, December 2005.

[23] A. Jbari, A. Adib, and D. Aboutajdine, "Blind source separation based on wavelet signal representation," in *Proceedings of the IEEE International Conference on Electronics, Circuits and Systems, ICECS 2007*, Marrakech, Morocco, December 2007, pp. 1155–1158.

[24] B. V. Gowreesunker and A. H. Tewfik, "Dictionary and sparse decomposition method selection for underdetermined blind source separation," in *Proceedings of the 15th European Signal Processing Conference EUSIPCO 2007*, Poznań, Poland, September 2007, pp. 768–772.

[25] R. Gribonval, "Sparse decomposition of stereo signals with matching pursuit and application to blind separation of more than two sources from a stereo mixture," in *Proceedings of the International Conference on Audio, Speech and Signal Processing, ICASSP 2002*, Orlando, Florida, May 2002.

[26] N. Mitianoudis and T. Stathaki, "Batch and online underdetermined source separation using laplacian mixture models," *IEEE Transactions on Audio, Speech and Language Processing*, vol. 15, no. 6, pp. 1818–1832, 2007.

[27] P. Bofill and M. Zibulevski, "Underdetermined blind source separation using sparse representations," *Signal Processing*, vol. 81, p. 23532362, 2001.

[28] O. Yilmaz and S. Rickard, "Blind separation of speech mixtures via time-frequency masking," *IEEE Transactions on Signal Processing*, vol. 52, no. 7, pp. 1830–1847, July 2004.

[29] S. Rickard, *Blind Speech Separation.* Springer, 2007, ch. 8: The DUET Algorithm.

[30] T. Virtanen, "Sound source separation in monaural music signals," Ph.D. dissertation, Tampere University of Technology, Finland, November 2006.

[31] R. Meddis, "Simulation of mechanical to neural transduction in the auditory perception," *Journal of the Acoustical Society of America*, vol. 79, no. 3, pp. 702–711, 1986.

[32] M. Cobos and J. J. Lopez, "Resynthesis of sound scenes in wave-field synthesis from stereo mixtures using sound source separation algorithms," *Journal of the Audio Engineering Society*, vol. 57, no. 3, March 2009.

[33] M. Vinyes, J. Bonada, and A. Loscos, "Demixing commercial music productions via human-assisted time-frequency masking," in *Audio Engineering Society 120th Convention*, Paris, France, May 2006.

In: Music: Composition, Interpretation and Effects
Editor: Tamara E. Ivanova, pp. 159-190
ISBN 978-1-60876-170-8

Chapter 10

COPYRIGHT PROTECTION OF MUSIC USING DIGITAL WATERMARKING BASED ON ECHO EFFECT

Nguyen T.H. Lien[1]*, ***Fangyan Dong***[1]†, ***Hajime Nobuhara***[2]‡
and Kaoru Hirota[1]§
[1]Tokyo Institute of Technology
[2]University of Tsukuba

Abstract

A cocktail digital watermarking scheme based on echo effect is proposed to embed copyright information into music clips before distribution and to retrieve this information later to prove the ownership. The proposed scheme takes advantage of stereo music to increase retrieval capacity and robustness by embedding data into the left and the right channel using low-bit coding and echo hiding method which have complimentary characteristics against various attacks. Furthermore, an exponential time-spread echo kernel is proposed for echo hiding to embed data as echoes into music clips while preserving their sound natural. By spreading the echo so that it seems like natural room reverberation, through the use of two channels, the proposed scheme provides satisfactory sound quality and exhibits both efficient retrieval capacity and robustness against attacks. Experiments show that the proposed system retrieves watermarks with 100% accuracy for all clips. The embedding transparency of the proposed scheme is -1.28 in Objective Different Grade (ODG) which is equal to "almost imperceptible". With a set of 42 attacks supplied by Stirmark Benchmark for Audio (SMBA), the proposed scheme has an average robustness of 0.98 where the ideal value is 1.00, and higher than the conventional methods.It also shows the robustness against mp3 compression

*E-mail address: nhlien@hrt.dis.titech.ac.jp
†E-mail address: tou@hrt.dis.titech.ac.jp
‡E-mail address: nobuhara@iit.tsukuba.ac.jp
§E-mail address: hirota@hrt.dis.titech.ac.jp

from 192kbps to 128 kbps. With these results, the proposed scheme is shown to be applicable to music distribution.

1. Introduction

Recently, there has been an increasing need for Digital Watermarking (DW) techniques. This is motivated by the ease of copying and manipulating digital media contents, especially over the internet. DW is a technique for the imperceptible insertion of secret copyright or other information into the media content. For music files, the embedded data should be inaudible to maintain the quality of the music (transparency), retrievable to confirm the authorship (retrieval capacity) and robust enough to withstand attack attempts such as removal or alteration of the embedded watermarks (robustness). DW, however, does not restrict access to the data. A watermark is permanently reside in the host data. When the ownership of a digital work is in question, the information can be extracted to completely characterize the owner. DW is especially useful for artists or companies that distribute music through Internet, as well as for music buyers since the former can prove the ownership of the music files, whereas the latter can prove their right of using those files.

A cocktail watermarking scheme for digital images is proposed [20], in which two modulation rules are used to hide two complementary watermarks in a host image so that at least one watermark survives under different types of attack. The cocktail watermarking method is proved to be remarkably effective in resisting various attacks.

In audio watermarking, there are many individual DW techniques based on the time domain [2], the frequency domain [3], [4] or the cepstrum domain [5] [6], wavelet domain [29]. It is difficult for an individual method to be robust against all attacks [26], [28]. No attention, however, has been given to a cocktail watermarking scheme for audio which uses more than one method to resist against different types of attacks. A two channel DW system which individually embeds the left and right channels using different methods is proposed. The DW methods used in this cocktail scheme must be complementary to each other in terms of robustness to various attacks, which is why the Least Significant Bit method [7] and Echo Hiding [8] are chosen for this study. The LSB method is a method to embed watermark into the bit which plays the least important role in quantization expression of samples and it is robust against attacks which do not change the positions of the samples. EH is a method which takes advantage of echo effects to embed watermark as echoes and it is strong against attacks which manipulate on the magnitude of the samples. Furthermore, both the LSB method and EH seldom generate noticeable noise and do not require the original signal during watermark decoding [9].

Recently, some improvements which use different echo effects for echo hiding such as negative single echo kernel [10], dual kernels [11], multiple echo kernels [11], backward and forward kernels [12] have been proposed to increase the robustness of embedded watermarks or the quality of watermarked clips. Oh et al. [10] proposed the negative echo hiding which can considerably reduce the noise perceptibility.However, their scheme suffers from low decoding rate. Kim et al. [12] proposed to embed watermarks with backward and forward kernels. The pair of backward and forward kernels helps to make the cepstrum peak more distinguishable, thus increasing the accuracy of the decoding process considerably. The sound quality, however, is slightly degraded. Recently, Ko et al. [13] have proposed a new echo kernel using PN sequences which gives more natural sound quality. The idea is to use the watermark as a PN sequence to spread out the echo in the echo kernel and then convolve with the whole original signal. This scheme is applicable only to zero-bit watermarking schemes which are designed only to detect the presence or the absence of the watermark in the watermarked clip.

In this work, a two channel digital watermarking system which separately watermarks the left channel using the LSB method and the right channel using EH based on an Exponential Time-spread Echo Kernel (ETEK) is proposed for multi-bit watermarking. The ETEK for EH is proposed as an extension of the time-spread echo kernel or multi echo kernel in terms of adding an exponential rule for spreading out the echoes. The watermark bit 1 or 0 is encoded as 2 kernels having the same spreading length, but different time delays, and each watermark bit is embedded in each small segment of the signal by convoluting one of these two kernels with that segment. By making multiple echoes that appear like natural reverberation and spreading them out according to an exponential decaying rule, the proposed system makes the watermark more imperceptible while keeping the robustness against various attacks.

Effectiveness of the proposed ETEK is analyzed on 10 music clips (44.1 kHz, 16 bits, 30s $\tilde{9}$0s) belong to different genres. The detection accuracy and subjective sound quality evaluation by seven young people of the ETEK are compared with those of conventional single EH.

The experiments to evaluate performance of the proposed two channel system are done on 30 stereo music clips, encoded in mp3 format, sampled at 44.1 kHz, quantized with 16 bits, belonging to 10 different genres of music, and are approximately 30s in duration. All the music clips are provided by a Japanese company distributing music through internet. For a fair evaluation, a triangle evaluation framework is applied, in which tranparency, retrieval capacity and the robustness are vertices. The transparency of the proposed two channel ETEK system as well as other comparable methods is measured using Objective Difference Grade (ODG) criteria [14] and implemented using the open software EAQUAL [15]. Comparison of the retrieval capacity and the robustness of the proposed system with those of conventional methods is done based on the watermark bit decoding rate which accurately extracts watermark bits from the watermarked clips before and after being submitted to single 42 attacks provided by the SMBA evaluation suite version 1.3.1 [16]-[18].

Furthermore, some comments made by two music semi-professionals about sound quality of ten classical music clips which were embedded watermarks by the proposed two channel system bring deeper understanding about the behavior of musical instruments against digital audio watermarking.

The proposed two channel system based on ETEK is presented in section 2. The effectiveness of ETEK is analyzed in section 3. Experimental environment and triangle evaluation framework are explained in section 4. The results are shown in section 5., and comments from music semi-professionals and the evaluation results are discussed in section 6.

2. The Proposed Two Channel System Based on an Exponential Time-Spread Echo Kernel for Multi-Bit Watermarking

2.1. Two Channel Digital Watermarking System

A DW system which separately watermarks the signals of the left and right channels of a music clip is presented. To obtain improved robustness, the LSB method and ETEK based EH, which have non-overlapping robustness to different types of attacks, are applied. The overall system is illustrated in Fig. 1. Two watermarks w_1, w_2 are respectively embedded into signals x_1, x_2 of the left and right channels of the original signal using the LSB method and EH, resulting in the watermarked signals $x_1^{'}, x_2^{'}$. These two signals are then unified to obtain the overall watermarked x signal for publishing. Attackers may then alter $x^{'}$, producing new signals $x^{''} = \{x_1^{''}, x_2^{''}\}$. The watermarks w_1, w_2 are:

$$w_1[i_1] \in \{0,1\}\,; l_1 = 0, 1, ..., L_1 - 1, \tag{1}$$

$$w_2[i_2] \in \{0,1\}\,; l_2 = 0, 1, ..., L_2 - 1. \tag{2}$$

The original signal $x = \{x_1, x_2\}$ includes N quantized samples in the range $[-1, 1)$.

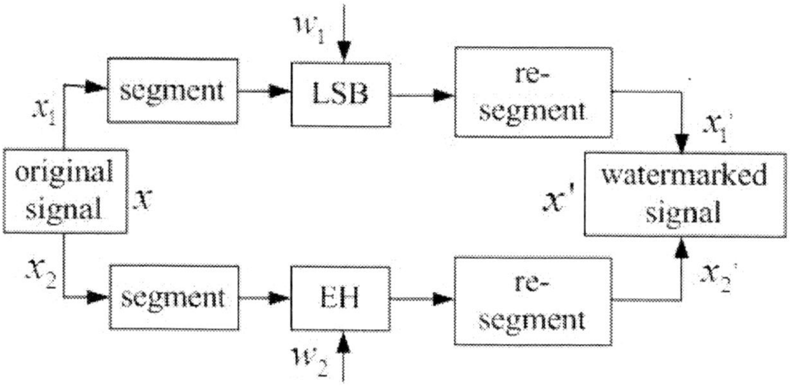

Figure 1. Two channel DW system based on the LSB method and EH with ETEK.

The original signal is divided into segments based on the watermark length. There is a portion of the signal which is not used for the watermarking process. Besides, signals from both channels are divided into segments such that each watermark bit is embedded in each segment. Therefore, the signals are considered as having the length $N = M_1L_1 = M_2L_2$, where M_1, M_2 are the number of samples in each segment, and L_1, L_2 are respectively the lengths of the watermark w_1, w_2 or the numbers of segments in the LSB channel and the ETEK based EH channel. The original signals of the two channels are represented by $x_1^{l_1}[i_1], x_2^{l_2}[i_2]$, where l_1, l_2 denotes the l_1^{th}, l_2^{th} segment, i_1, i_2 denotes the i_1^{th}, i_2^{th} sample in the segment. It is observed that in many clips, the amplitude of the right channel seems to be slightly larger than that of the left. For EH, the larger the amplitude is, the higher the extraction accuracy. For LSB, the accuracy of the extraction process does not depend on the signal amplitude, and applying it to either the left or the right channel does not change the result. Therefore, it is better to apply EH into the right channel.

To extract the watermarks from watermarked clips and attacked clips, two decoding methods are applied accordingly to the left and right channel. Then, the extracted watermarks are compared with the original watermark and the extracted watermark with highest accuracy is considered as the final result.

2.2. The LSB Method for the Left Channel

2.2.1. The Embedding Process

In the LSB channel, in each segment, one watermark bit is embedded at the i_1^{th} sample. The sample value is in $[-1, 1)$, with a 16-bit amplitude resolution, providing $2^{16} = 65536$ quantization levels:

$$x_1^{(l_1)}[i_1] \in \left\{ \left(\frac{k}{2^{15}} - 1 \right); k = 0, 1, 2, ..., (2^{16} - 1) \right\},$$

$$i_1 = 0, 1, ..., M_1 - 1, l_1 = 0, 1, ..., L_1 - 1,$$

$$x_1^{(l_1)}[i_1] = \left(\sum_{n=0}^{15} \frac{x_1^{(l_1)}[i_1]_n 2^n}{2^{15}} \right) - 1, \tag{3}$$

where $x_1^{(l_1)}[i_1]_n$ is the nth bit of $x_1^{(l_1)}[i_1]$ in 16-bit amplitude resolution, $x_1^{(l_1)}[i_1]_0$ is the most significant bit and $x_1^{(l_1)}[i_1]_{15}$ is the least significant bit.

The LSB method replaces the LSB bit with the watermark bit. In practice, embedding it at a bit position between the 9th and 15th bits does not affect the audio quality. Suppose that the watermark bit is embedded in the jth bit, where $j \in \{0, 1, ..., 15\}$, the embedding process is:

$$x_1^{'(l_1)}[i_1] = \left(\sum_{\substack{n=0 \\ n \neq j}}^{15} \frac{x_1^{(l_1)}[i_1]_n 2^n}{2^{15}} \right) + \frac{w_1[i_1] 2^j}{2^{15}} - 1. \tag{4}$$

2.2.2. The Decoding Process

The decoding process decodes the watermark bits by reading the jth bit values in the 16-bit expression of the i_1^{th} samples at each segment from the received signal:

$$x_1^{''(l_1)}[i_1] = \left(\sum_{\substack{n=0 \\ n \neq j}}^{15} \frac{x_1^{''(l_1)}[i_1]_n 2^n}{2^{15}} \right) - 1, \tag{5}$$

The decoded watermark is as follows:

$$w_1^{''(l_1)} = \left\{ x_1^{(l_1)}[i_1]_j; i_1 = 0, 1, ..., M_1 - 1 \right\}. \tag{6}$$

2.3. Exponential Time-Spread Echo Kernel based Echo Hiding for the Right Channel

2.3.1. Echo Hiding and Motivation for Exponential Time-Spread Echo Kernel

The EH embeds watermark bits in the original signal by introducing an "echo"[10]. The embedding system uses two echo kernels which are the same in magnitude but with different delays, one to represent a binary "1"and another to represent a binary "0". Both delays are below the threshold at which human ear can resolve the echo and the original signal as different sources. The decoding process calculates the cepstrums of the watermarked signal at two positions of delay one and zero, and compares them to decide which bit was embedded.

Fig. 2 shows the conventional embedding process using a single echo kernel. The embedding process can be represented as a system having two transfer functions as shown in Fig. 2(a). These two transfer functions are the echo kernels with two different delays: delay for bit 1 (Fig. 2(b)), and delay for bit 0 (Fig. 2(c)). If the echo amplitude is small, the method, however, suffers from low retrieval rate. In order to keep a sufficiently high retrieval rate, the echo amplitude should be increased. The problem is that the larger echo may cause an annoying perception of noise or bad quality sound. Thus, how to increase the echo amplitude while keeping the echo imperceptible is the purpose of improvements such as negative echo kernel [10], multiple echo kernel [11], and backward and forward echo kernel [12].

A single echo of a signal (as in Fig. 3(a)) happens when there is only a single wall between the sound source and the listener. In that case, the sound from the source to the receiver has two paths, one is directly from the source and the other is the reflected path from a single wall. In a real room, however, the sound pressure wave expands from the sound source (a point) radially outward, reaching walls and other objects where energy is both reflected and absorbed. It means that sounds in a real room includes both direct sound and many echoes heard as reverberation because of the multiple reflections from walls and other objects in the room. Moreover, the magnitude of the echoes reaching the listener varies according to the material of the walls and objects which reflect the echo, and the distance of the path that those echoes follow. If watermarks can be embedded with various displacements from the natural direct sound, such complicated echoes can provide a more natural sound quality than only a single echo even if those echoes are audible.

To improve the decoding rate, while keeping the sound quality natural, the ETEK shown in Fig. 3(b), which spreads out the echo according to an exponential decaying rule, is proposed based on the above room reverberation concept. The proposed ETEK is applied to the EH channel in the two channel DW system.

2.3.2. Watermark Embedding with Echo Hiding Based on an Exponential Time-Spread Echo Kernel

In EH, the signal is divided into L_2 segments, each segment includes M_2 samples:

$$x_2^{(l_2)}[i_2] \in [-1, 1)\,; i_2 = 0, 1, ..., M_2 - 1; l_2 = 0, 1, ..., L_2 - 1. \tag{7}$$

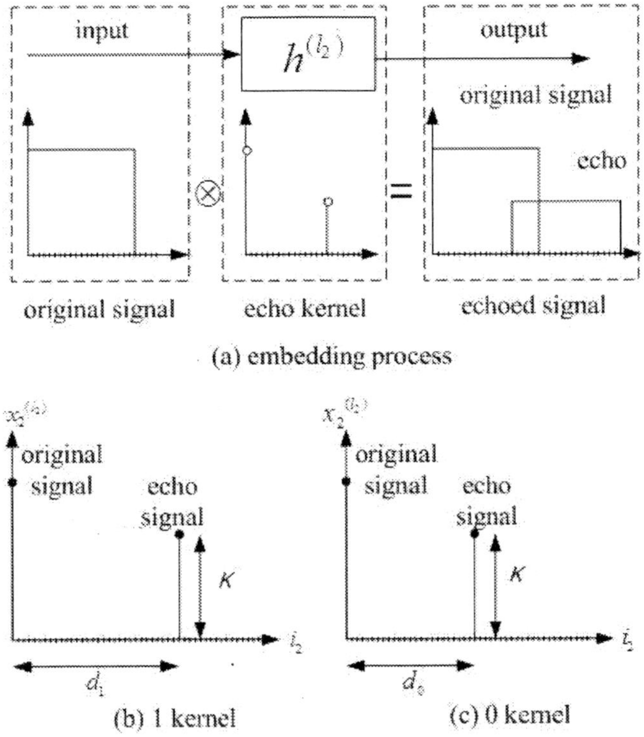

Figure 2. (a) Embedding process; (b) kernel 1 for embedding bit 1; (c) kernel 0 for embedding bit 0.

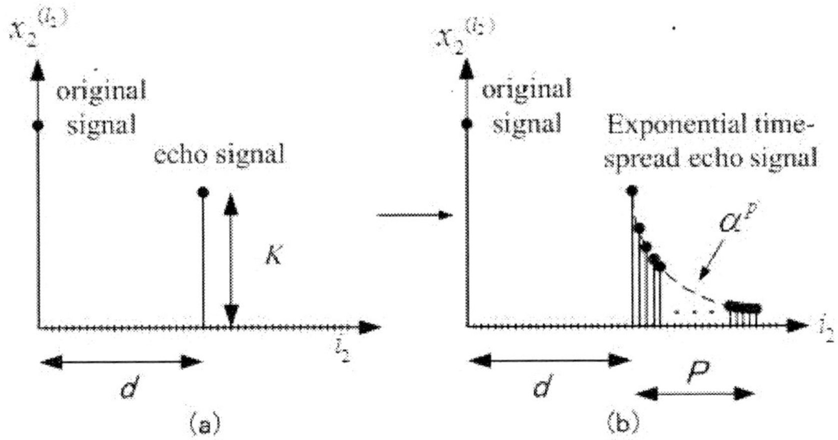

Figure 3. (a) Single Echo Kernel; (b) The proposed time-spread echo kernel.

Each watermark bit is embedded in each segment of the original signal.

The proposed ETEK in Fig. 3(b) is constructed from a decaying exponential function as:

$$h^{(l_2)}[i_2] = \delta^{(l_2)}[i_2] + \sum_{p=1}^{P} \alpha^p \delta^{(l_2)}[i_2 - d - p + 1]; -1 < \alpha < 1; P \leq L_2 - d, \quad (8)$$

where d is the delay and P is the number of echoes generated by the proposed kernel. These echoes, denoted by $K_i (i = 1, 2, ..., P)$, are spread over time with exponentially decreasing magnitudes, so that:

$$K_1[i_2] = \alpha^1 x_2^{(l_2)}[i_2], K_2[i_2] = \alpha^2 x_2^{(l_2)}[i_2], ..., K_p[i_2] = \alpha^p x_2^{(l_2)}[i_2]. \quad (9)$$

The number of echoes must be chosen so that the last echo K_P still belongs to the time range where the human auditory system can not distinguish the audio signal from the echo K_P (approximately 2ms).

Fig. 4 shows a schematic explanation of how data is embedded into the original signal by the ETEK. To embed the watermark bit 1 and 0, the embedding process uses two kernels similar in spread length and different in the delay d of the first echo. The echo kernels for embedding bit 0 and bit 1 respectively are as follows:

$$h^{(l_2)}[i_2] = \delta^{(l_2)}[i_2] + \sum_{p=1}^{P} \alpha^p \delta^{(l_2)}[i_2 - d_0 - p + 1]; -1 < \alpha < 1; P \leq L_2 - d, \quad (10)$$

$$h^{(l_2)}[i_2] = \delta^{(l_2)}[i_2] + \sum_{p=1}^{P} \alpha^p \delta^{(l_2)}[i_2 - d_1 - p + 1]; -1 < \alpha < 1; P \leq L_2 - d. \quad (11)$$

The watermarked signal is obtained by taking a linear convolution of the original signal and the ETEK:

$$\begin{aligned} x_2^{'(l_2)}[i_2] &= x_2^{(l_2)}[i_2] \otimes h^{(l_2)}[i_2] \\ &= x_2^{(l_2)}[i_2] \otimes \left(\delta^{(l_2)}[i_2] + \sum_{p=1}^{P} \alpha^p \delta^{(l_2)}[i_2 - d - p + 1] \right). \end{aligned} \quad (12)$$

In the other words, the watermarked signal is composed of the original signal and P echoes generated by convoluting the ETEK with the original signal.

2.3.3. The Decoding Process

In EH, the cepstrum function of the watermarked signal $x_2^{'(l_2)}$ plays an essential role in finding out which echo kernel (echo kernel for bit 0 or bit 1) was used in each segment of the signal, and then the watermark bits are decoded accordingly. The cepstrum function is defined as:

$$C_{x'^{(l)}}[i] = F^{-1}_{x'^{(l)}} \left(\{ln F_{x'^{(l)}}[\varpi]\}_{\varpi=0}^{L-1} \right) [i] \quad (13)$$

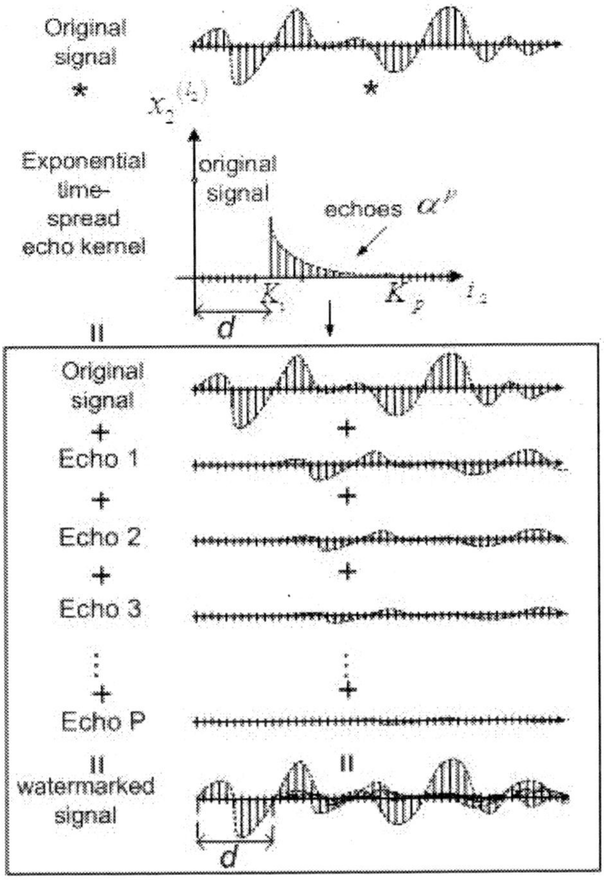

Figure 4. Watermark embedding process based on exponential time-spread echo kernel.

where F and F^{-1} denote the Fourier transform and the inverse Fourier transform, respectively [22] [30]. The Fourier transform is:

$$F_{x'^{(l)}}[\varpi] = \sum_{i=0}^{L-1} x'^{(l)}[i] e^{-j\left(\frac{2\pi\varpi}{L}\right)}, \tag{14}$$

and $\{lnF_{x'^{(l)}}[\varpi]\}_{\varpi=0}^{L-1}$ is defined as:

$$\{\ln F_{x'^{(l)}}[\varpi]\}_{\varpi=0}^{L-1} = \{\ln F_{x'^{(l)}}[\varpi] \in C \,|\varpi = 0, 1, 2, ..., L-1\}, \tag{15}$$

where C is the set of complex numbers. The inverse Fourier transform of equation 15 is:

$$F_{x'^{(l)}}^{-1}\left(\{\ln F_{x'^{(l)}}[\varpi]\}_{\varpi=0}^{L-1}\right)[i] = \frac{1}{L} \sum_{\varpi=0}^{L-1} \ln\left(F_{x^{(l)}}\right) e^{j\left(\frac{2\pi\varpi}{L}\right)i}. \tag{16}$$

Then, the cepstrum of $x'^{(l)}$ can be rewritten as:

$$\begin{aligned} C_{x'^{(l)}}[i] &= F^{-1}_{x^{(l)}\otimes h^{(l)}}\left(\{\ln F_{x^{(l)}\otimes h^{(l)}}[\varpi]\}_{\varpi=0}^{L-1}\right)[i] \\ &= F^{-1}_{x^{(l)}\otimes h^{(l)}}\left(\{\ln F_{x^{(l)}}[\varpi].F_{h^{(l)}}[\varpi]\}_{\varpi=0}^{L-1}\right)[i] \\ &= F^{-1}_{x^{(l)}\otimes h^{(l)}}\left(\{\ln F_{x^{(l)}}[\varpi]\}_{\varpi=0}^{L-1} + \{\ln F_{h^{(l)}}[\varpi]\}_{\varpi=0}^{L-1}\right)[i] \\ &= F^{-1}_{x^{(l)}}\left(\{\ln F_{x^{(l)}}[\varpi]\}_{\varpi=0}^{L-1}\right)[i] + F^{-1}_{h^{(l)}}\left(\{\ln F_{h^{(l)}}[\varpi]\}_{\varpi=0}^{L-1}\right)[i] \end{aligned} \quad (17)$$

Equation 17 can be rewritten as a separable cepstrum of original signal component $x^{(l)}$ and echo components $h^{(l)}$ as follows:

$$C_{x_2'^{(l_2)}}[i_2] = C_{x_2^{(l_2)}}[i_2] + C_{h^{(l_2)}}[i_2], \quad (18)$$

with

$$C_{h^{(l_2)}}[i_2] = F^{-1}_{h^{(l_2)}}\left(ln\left\{F_{h^{(l_2)}}[w]\right\}_{w=0}^{L_2-1}\right)[i_2], \quad (19)$$

where F and F^{-1} denote the Fourier transform and the inverse Fourier transform of the echo kernel when $w = 0, ..., L_2 - 1$, respectively. Applying ceptrum analysis for $C_{h^{(l_2)}}[i_2]$, we get:

$$\begin{aligned} C_{x_2'^{(l_2)}}[i_2] = C_{x_2^{(l_2)}}[i_2] + \alpha\left(\delta[i_2 - d] + ... + \delta[i_2 - d - P]\right) \\ -\frac{\alpha^2}{2}\left(\delta[i_2 - 2d] + \delta[i_2 - 2d - 2] + ...\right) + ... \end{aligned} \quad (20)$$

The cepstrum of $C_{x_2'^{(l_2)}}[i_2]$ at time d is:

$$C_{x_2'^{(l_2)}}[d] = C_{x_2^{(l_2)}}[d] + \alpha. \quad (21)$$

If the embedded watermark bit is 1, the cepstrum of $C_{x_2'^{(l_2)}}[i_2]$ at delay d_1 is $C_{x_2'^{(l_2)}}[d_1] = C_{x_2^{(l_2)}}[d_1] + \alpha$ and at delay d_0 is just $C_{x_2'^{(l_2)}}[d_0] = C_{x_2^{(l_2)}}[d_0]$; and if the embedded watermark bit is 0, the cepstrum of $C_{x_2'^{(l_2)}}[i_2]$ at d_0 is $C_{x_2'^{(l_2)}}[d_0] = C_{x_2^{(l_2)}}[d_0] + \alpha$ and at d_1 is $C_{x_2'^{(l_2)}}[d_1] = C_{x_2^{(l_2)}}[d_1]$.

To decode watermark bits, the signal in the ETEK channel is first segmented, and for each segment, the cepstrum of that signal at delay d_0 and d_1 are calculated. The watermark bit for the segment l_2 is decoded as:

$$w_2''^{(l_2)} = \begin{cases} 1 & \text{if } C_{x_2'^{(l_2)}}[d_1] > C_{x_2'^{(l_2)}}[d_0] \\ 0 & \text{if else} \end{cases} \quad (22)$$

3. Analyzing the Effectiveness of ETEK

3.1. Audio Data

In order to analyze the effectiveness of ETEK independently, a set of 10 stereo music clips (44.1 kHz, 16 bits, 30s 90s) is applied and used to compare the proposed ETEK-EH with conventional single echo hiding.

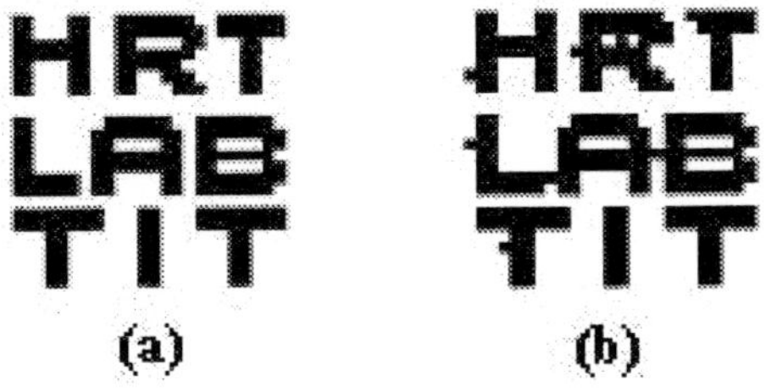

Figure 5. (a) Embedded logo image, (b) extracted logo image by echo hiding.

3.2. Authorized Data for Embedding

A 26x26 binary logo image (676 bits) shown in fig.5(a) is used as watermark. With such a watermark, more robustness can be obtained because the logo can always preserve a certain degree of structural information, which is meaningful and recognizable. The extracted watermark is easier to verify with human eyes rather than by the correlation method. Producer, company and users can insert their ones instead of that logo.

3.3. Parameters

The logo image as in fig.5(a) is embedded into audio music clips by the conventional echo hiding method, and the proposed exponential time-spread echo kernel method. The delay d_1 for embedding bit 1 is set to $d_1 = 1.3ms$, the delay d_0 for embedding bit 0 is set to $d_0 = 1.0ms$. The parameter α is set variously from 0.1 to 0.5 to check the variation of the results depending on it.

Fig.5(b) shows the image logo which is extracted from the conventional EH channel of one embedded clip. The extracted logo image show that EH can correctly extract the embedded image and can exactly specify the author of the audio content. This result also proves the advantage of logo image compared to other kinds of data such as pseudo-random sequence and text data. The pseudo-random sequence data can not be verified by human eyes, and the text can not be recovered back to the original text form if the data extracted rate is not 100% whereas the logo image can be recovered and verified by human eyes even when the watermark extracted rate is not 100%.

3.4. Sound Quality Evaluation

The quality of one clip embedded by conventional single echo hiding and the proposed ETEK method is subjectively evaluated by seven people in their twenty with normal hearing. The listeners are asked to compare the watermarked clip with the original clip in terms of the listening quality and to give a grade according to following criterion: Imperceptible = 5, perceptible but not annoying = 4, slightly annoying = 3, annoying = 2, very annoying = 1. And the average results are 4.57 for the clip which is applied to the proposed ETEK, and 3.86 for the one applied to the conventional single EH.

All the watermarked clips which are applied to the proposed ETEK are listened and evaluated by an artist. With the same parameters, sound quality of different embedded clip is different. For Pop and Rock, it is easier than other genres to be recognized the difference on drum sounds. Besides, high vocal sound of female singer is easy to be degraded (Clip 1, 2). To be able to be applied to the website distributing music, which requires that all the watermarked clips can not be recognized the differences by every listener, "the parameter should be set in the ranges of 0.2 to 0.25."

3.5. Accuracy of the Detection Process

The accuracy of the detection process is calculated based on the number of the extracted bits correctly identified. Let's take A to symbol the accuracy rate of detection process. That is:

$$A(\%) = \frac{100}{M} \sum_{i=0}^{M-1} \begin{cases} 1, & w[i] = w''[i] \\ 0, & w[i] \neq w''[i] \end{cases} \tag{23}$$

The conventional EH and the proposed ETEK are applied to ten clips. The echo rate α for both methods varies from 0.1 to 0.4. The sound of the clips which are watermarked by the ETEK is more natural when P increases, however, the calculation time increases. Actually, when $P > 5$ $(\alpha \leq 0.5)$, the rate of following echoes becomes smaller than $\frac{1}{10}$ times of the first echo, small enough to be not taken into account. The number of echoes P is set to 5.

The accuracy varies according to clip. For pop or rock clips, where magnitude of the sound is comparatively large and the silence interval is few, the accuracy of the detection is high. For ambient or semi-classic clips, where the magnitude of the sound is small and there are more silence intervals, the accuracy of the detection is low. Therefore, clips with genre such as pop, rock, and gospel are more robust than clips with genre such as ambient, semi-classic.

Table 1 shows the accuracy comparisons of the single echo kernel and the proposed ETEK in the ten clips when the parameter varies from 0.1 to 0.4. Fig.6 also indicates accuracy comparison of two methods in case of the clip 6. The proposed echo kernel gives higher detection rate for all α. The experiments on different clips also show that the proposed echo kernel yields higher accuracy rate for all music clips and parameter α. Especially, for ambient and semi-classic genres which have low detection rate in the conventional EH, the proposed echo kernel helps to increase the detection rate very much. For example, the detection rate of clip 7 (Ambient) increases 14% when $\alpha = 0.4$. For Pop and Rock genres, the proposed echo kernel also helps to increase the detection rate, but not as much as ambient or semi-classic.

Fig.7 shows the accuracy comparisons of the conventional kernel and the proposed kernel when $\alpha = 0.25$ for 10 music clips. Experiments confirm that the watermark detection rate becomes high as the parameter α increases. If the parameter α increases, however, the original music is degraded. In the point of view of sound quality, the parameter α should be small but in the point of view of robustness, the parameter α should be large. For music, the most important feature is the sound quality, and then the robustness. Experiments show that the proposed ETEK give higher detection rate than the conventional single echo kernel. It means that, with the same detection rate, the parameter α for ETEK is smaller than that for the conventional echo kernel, or with the same parameter , the proposed ETEK gains higher sound quality.

Table 1. Accuracy comparision of the conventional EH and the proposed ETEK (%)

α		1	2	3	4	5	6	7	8	9	10
0.10	EH	48.81	48.07	68.2	69.23	75.73	59.91	50.59	42.15	67.01	45.25
	ETEK	49.26	49.26	69.08	70.11	77.07	61.39	51.00	42.45	68.49	44.67
0.15	EH	48.81	55.17	79.14	78.84	84.76	68.19	52.36	43.63	75.44	45.85
	ETEK	53.99	59.61	79.58	79.73	86.24	73.81	53.55	44.82	76.77	46.15
0.20	EH	54.28	63.90	87.42	78.84	90.68	77.66	53.69	49.85	80.62	48.37
	ETEK	59.91	67.00	87.42	87.72	91.56	80.62	56.21	50.59	82.10	50.59
0.25	EH	61.98	72.18	91.42	91.42	93.49	81.95	56.65	54.73	84.76	49.85
	ETEK	65.23	77.21	92.6	92.30	93.93	84.76	62.72	58.57	87.57	53.84
0.30	EH	66.71	77.81	92.89	91.42	94.82	81.95	59.61	63.46	88.60	53.84
	ETEK	75.73	84.17	93.78	95.26	95.56	88.75	64.64	68.78	91.27	60.65
0.35	EH	72.18	82.10	94.82	96.00	95.56	86.39	60.50	73.66	91.27	60.94
	ETEK	80.91	88.31	95.71	95.85	96.00	90.38	71.30	77.95	92.60	66.12
0.40	EH	75.59	86.53	95.85	97.63	95.71	88.16	62.86	77.07	91.86	62.86
	ETEK	85.20	91.27	96.89	97.18	96.15	91.86	76.92	82.39	93.93	69.82

4. Experimental Environment and Triangle Evaluation for Digital Watermarking

Transparency, capacity and robustness are three trade-off properties that almost all DW systems consider. To get an objective evaluation and comparison of performance between the proposed system and conventional methods, the experiments follow the theoretical evaluation framework introduced by J. Dittmann et al., 2006 [19].

4.1. Music Data and Watermark

All 6 watermarking systems use the same audio test set S, which contains 30 stereo music clips, supplied by a Japanese company distributing music over the Internet. All music clips are sampled at 44.1 kHz, quantized by 16 bits, approximately 30s of duration, encoded in mp3 format in 192kbps and belong to 10 different genres including 9 clips of Pop, 5 clips of Rock, 2 clips of Jazz, 3 clips of Healing/Ambient, 2 clips of Folk, 2 clips of Techno/House, 3 clips of Gospel, 1 clip of Blues, 1 clip of Semi-classic, 2 clips of World music among others.

To make the evaluation process simpler, the proposed system and the other comparative methods use the same watermark w of 676 bits as in Fig. 5(a). All 30 clips $s \in S$ are watermarked with the 6 different watermarking systems $\Omega = \{\Omega_k, k = 1, ..., 6\}$ to get 6 watermarked clip sets $S(\Omega_k, k = 1, ..., 6)$ (180 watermarked clips overall). To evaluate the performance of DW systems, these watermarked clip sets are run through the retrieval process R, and submitted to a set A of 42 attacks provided by the SMBA to get the attacked clip set $S(\Omega_k A_n, k = 1, ..., 6, n = 1, ..., 42)$ (7560 attacked clips in overall).

Besides, in order to make easier to analyze the relation between different measurements and compare different methods, all the measurements (except the embedding capacity) are normalized to the range $[0, 1]$.

4.2. Comparative Methods: Basic Definitions and Parameter Setting

The proposed system (Ω_6) is compared with the LSB method (Ω_1), EH using different echo kernels: single echo kernel [10] (Ω_2), backward and forward echo kernel [12] (Ω_3), multiple echo kernel [11] (Ω_4), and negative echo kernel [10] (Ω_5). The embedding methods $E = \{E_k, k = 1, ..., 6\}$ and retrieval methods $R = \{R_k, k = 1, ..., 6\}$ of the above 6 DW systems $\Omega = \{\Omega_k, k = 1, ..., 6\}$ with their parameters are described below.

4.2.1. Least Significant Bit

The LSB method embeds the watermark bits in the bit that is least significant of the host signal [7]. In the host signal that is quantized by 16 bit, the 16th bit is the LSB. This method is very simple but give a high embedding capacity.

The LSB method Ω_1 is applied to both channels with the same parameters: $j_1 = 9, i_1 = i_2 = 100$. (The retrieval function will retrieve the watermark from two channels of the test signal and take the best result as the final result).

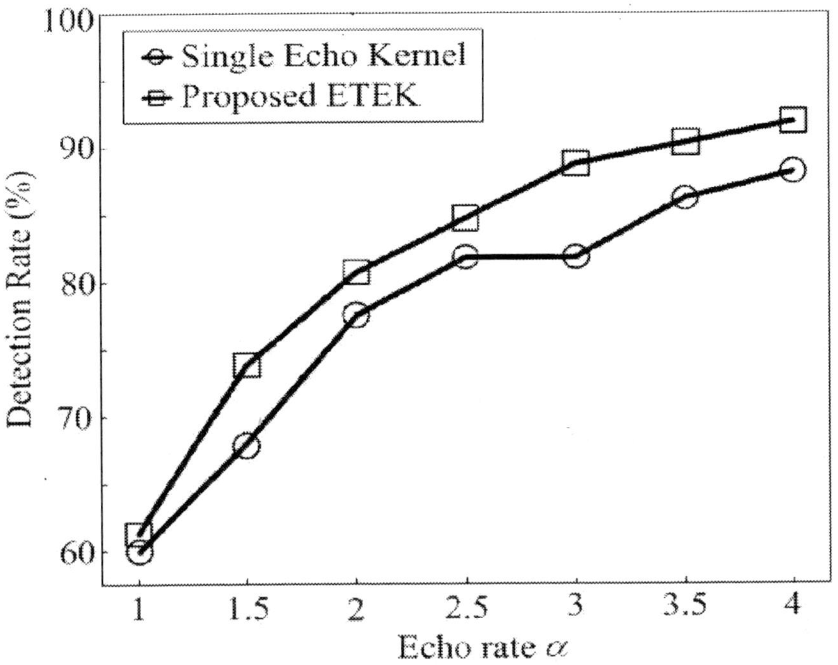

Figure 6. Accuracy comparisons of the conventional EH and the proposed ETEK of Clip 6 when α is in the range of 0.1 to 0.4.

4.2.2. Single Echo Hiding

This is the first echo hiding method where the system functions include only a single positive pulse to present for the echo [8], as shown in Fig.8(a).

Single EH Ω_2 is applied to both channels with the echo delay for bit 1 $d_1 = 1.0ms$, the echo delay for bit 0 $d_0 = 1.3ms$, the echo amplitude $\alpha = 0.4$, and the number of echoes $P = 1$.

4.2.3. Multiple Echo Hiding

Multiple EH is an improvement of single EH where the system functions include multiple positive pulses to present for multiple echoes, as shown in Fig.8(b).

Multiple EH Ω_3 is applied to both channels with the number of echoes $P = 4$ and the same amplitude $\alpha = 0.4$ for all echoes. The echo delay for bit 1 $d_1 = 1.0ms$, and the echo delay for bit 0 $d_0 = 1.3ms$.

4.2.4. Negative Echo Hiding

This is also another improvement of single EH. It is, however, done in the fashion that decrease the amplitude of the host signal, instead of adding more sounds like other EH methods, by using negative pulse in the system functions, as shown in Fig.8(c).

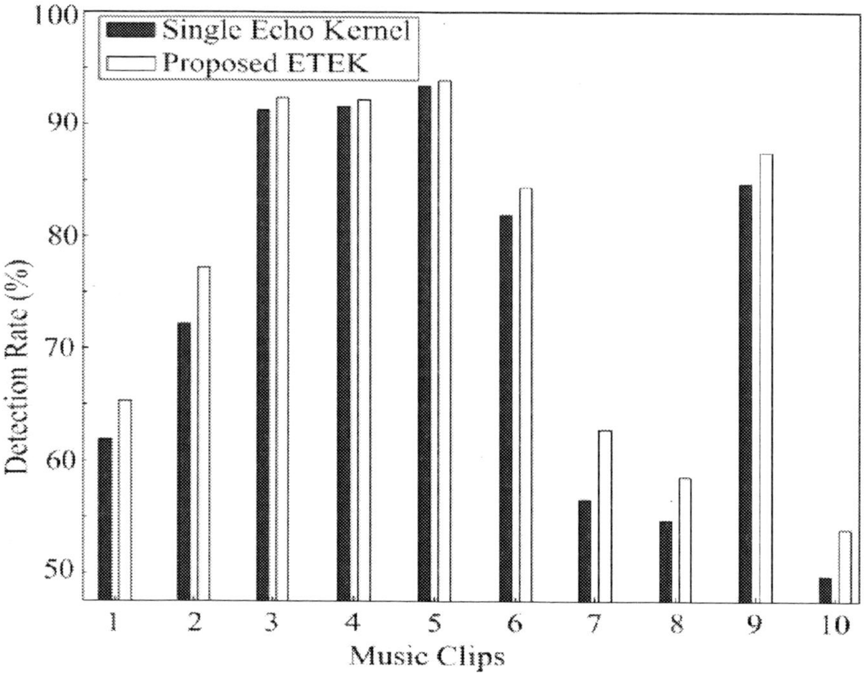

Figure 7. Accuracy comparisons of the conventional EH and the proposed ETEK of Clip 10 when $\alpha = 0.25$.

Negative EH Ω_4 is applied to both channels with echo amplitude $\alpha = 0.4$, echo delay for bit 1 $d_1 = 1.0ms$, echo delay for bit 0 $d_0 = 1.3ms$, and the number of echoes $P = 1$.

4.2.5. Backward and Forward Echo Hiding

This is a compatitive improvement of EH which makes use of the characteristic of cepstrum analysis for the bilateral pulse of system functions, as shown in Fig.8(d).

Backward and Forward EH Ω_5 is applied to both channels with the number of echoes $P = 2$. The backward echo and forward echo delays for bit 1 are $d_1 = 1.0ms$, echo delays for bit 0 are $d_0 = 1.3ms$, and the echo amplitude $\alpha = 0.4$.

4.2.6. The Proposed Two Channel Digital Watermarking

The proposed system Ω_6 with the left channel is watermarked using the LSB method, and the right channel is watermarked with EH using the proposed ETEK. For the LSB channel, each watermark bit is embedded in the j_1^{th} bit position $(j_1 = 9)$ of the i_1^{th} sample $(i_1 = 100)$ of each segment. For the ETEK channel, the echo delay for bit 1 $d_1 = 1.0ms$, the echo delay for bit 0 $d_0 = 1.3ms$, the echo amplitude $\alpha = 0.4$, and the number of echoes $P = 5$, spread out following the rule (α^p). The number of echoes P is set to 5 because when $P > 5(\alpha < 0.5)$, the amplitude of the following echoes becomes smaller than 1/10 times of the first echo, small enough to be ignored.

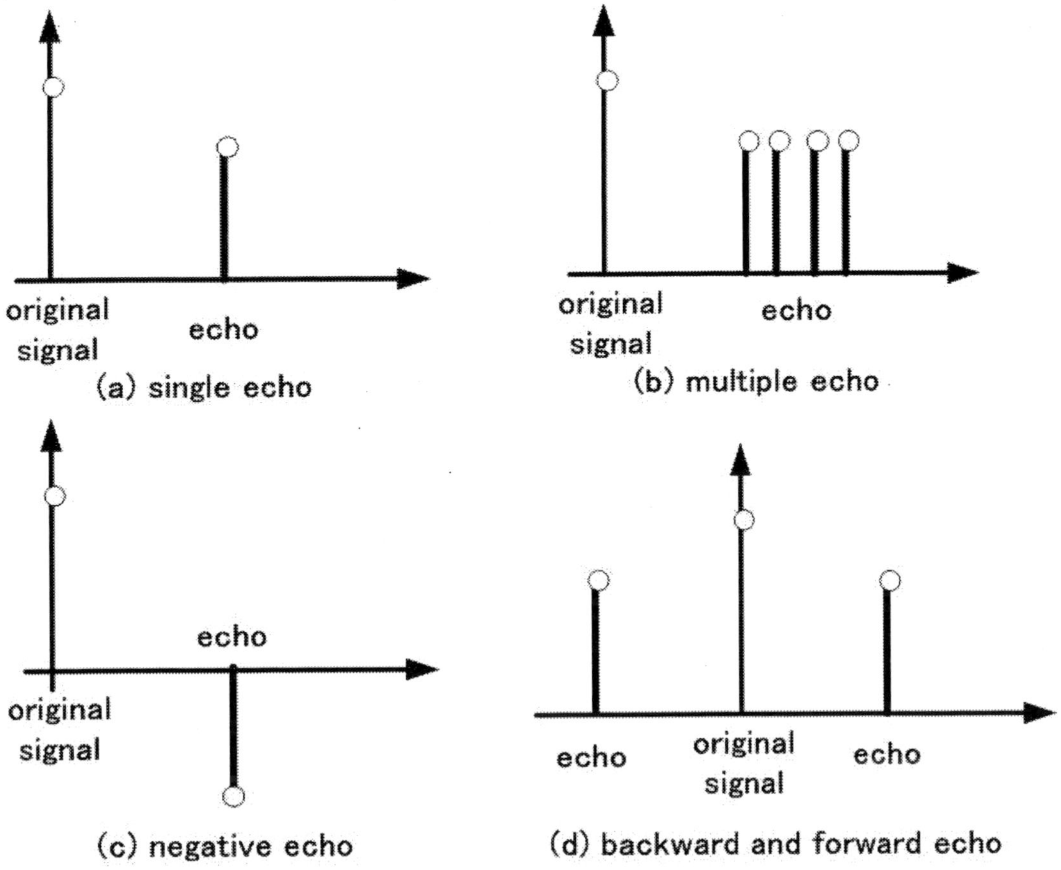

Figure 8. Comparative echo hiding methods.

All the above embedding parameters are also used to extract the watermarks and summarized in 2.

4.3. The Embedding Process

For the embedding process, the number of bits of the watermark that were embedded, e.g. the capacity of the embedding function cap_E, and the sound quality of the watermarked clip, e.g. the transparency tra_E, are the parameters that need to be measured.

Transparency: The transparency relates to the degree of distortion introduced by the embedding function of a watermarking system. It is defined in an interval $[0, 1]$, where 0 indicates the worst case (the watermarked clip was distorted so much that it is considered totally different from the original clip) and 1 is the best case (the listener does not perceive any difference between the two clips). In this work, the ODG (Objective Difference Grade) value is used to measure how different the watermarked clip is from the original clip, which is in the range $[-4, 0]$ (where -4 is perceptible and very annoying, -3 is perceptible and slightly annoying, -2 is perceptible but not annoying, -1 is almost imperceptible and 0 is completely imperceptible) [14]. Then, the ODG values are scaled to $[0, 1]$ range by com-

Table 2. Embedding parameters of 6 systems

Ω	Embedding Parameters
Ω_1	$j_1 = 9, i_1 = i_2 = 100$
Ω_2	$d_1 = 1.0ms, d_0 = 1.3ms, \alpha = 0.4, P = 1$
Ω_3	$d_1 = 1.0ms, d_0 = 1.3ms, \alpha = 0.4, P = 4$
Ω_4	$d_1 = 1.0ms, d_0 = 1.3ms, \alpha = -0.4, P = 1$
Ω_5	$d_1^b = -1.0ms, d_1^f = 1.0ms, d_0^b = 1.3ms, d_0^f = 1.3ms, \alpha = 0.4, P = 2$
Ω_6	$j_1 = 9, i_1 = i_2 = 100, d_1 = 1.0ms, d_0 = 1.3ms, \alpha = 0.4, P = 5$

puting
$\left(1 - \frac{|ODG|}{4}\right) \rightarrow [0, 1]$ and this value is considered to be the relative transparency of a watermarking system for a specific clip tra_{Erel}^s. To evaluate the transparency of a watermarking system for an overall data set S, the average, maximal and minimal transparency of E for data set S is measured $(tra_{Eave}, tra_{Emax}, tra_{Emin})$:

$$\begin{aligned} tra_{Eave} &= \frac{\sum_{s=1}^{30} tra_{Erel}^s}{30}, \\ tra_{Emax} &= max_{s=1}^{s=30}(tra_{Erel}^s), \\ tra_{Emin} &= max_{s=1}^{s=30}(tra_{Erel}^s). \end{aligned} \tag{24}$$

Capacity cap_E: is the amount of information that is embedded in the clip and relative to the length of the original audio signal. In the experiments, because all music clips have the same length (30s) and all the 6 watermarking systems use the same watermark (676 bits) for all the 30 clips, the embedding capacity is considered as $cap_E = 676$.

4.4. The Retrieval Process

Retrieval capacity: is the rate between the watermark bits correctly retrieved from the watermarked clip and the original watermark bits:

$$cap_{Rrel}^s = \frac{1}{676} max_{i(i=i_1,i_2)} \left(\sum_{i=1}^{676} \begin{cases} 1 & \text{if } w[i] = w_R[i] \\ 0 & \text{otherwise} \end{cases} \right), \tag{25}$$

where $w[i], w_R[i], i = 1, 2, ..., 676$ are the embedded and the retrieved watermark.

To evaluate the retrieval capacity for an overall data set S, the average, maximal and minimal capacity for S are measured $(cap_{Rave}, cap_{Rmax}, cap_{Rmin})$:

$$\begin{aligned} cap_{Eave} &= \frac{\sum_{s=1}^{30} cap_{Erel}^s}{30}, \\ cap_{Emax} &= max_{s=1}^{s=30}(cap_{Erel}^s), \\ cap_{Emin} &= max_{s=1}^{s=30}(cap_{Erel}^s). \end{aligned} \tag{26}$$

The detection success rate det_R^s is defined as:

$$det_{R\lambda}^s = \begin{cases} 1 & \text{if } cap_{Rrel}^s \\ 0 & \text{otherwise} \end{cases} \quad (27)$$

The threshold λ is set to 0.7. This means that correctly retrieving more than 70% of the watermark is considered to be a success attempt.

To measure the detection success over a test set, the average of det_R can be computed as:

$$det_{R\lambda ave} = \frac{1}{30}\sum_{s=1}^{30} det_{R\lambda}^s. \quad (28)$$

4.5. The Attacking Process

For each watermarking system, we produce the set S of 30 watermarked clips. Each watermarked clip is submitted to 42 attacks with default parameters provided by SMBA [16]-[18], producing 42 attacked clips. Finally, the attacked clip set S' of the watermarked clip set includes 1260 attacked clips for each watermarking system (7560 attacked clips overall for the 6 systems).

Capacity: the capacity $cap_{Areal}^{S'}$ of an attacked clip is the rate between watermark bits correctly retrieved from the attacked clip and the watermark bits:

$$cap_{Arel}^{S'} = \frac{1}{676} max_{i(i=i_1,i_2)}\left(\sum_{i=1}^{676}\begin{cases} 1 & \text{if } w[i] = w_A[i] \\ 0 & \text{otherwise} \end{cases}\right), \quad (29)$$

where $w[i], w_A[i], i = 1, 2, ..., 676$ are the embedded watermark and the retrieved watermark after being attacked.

To evaluate the retrieval capacity for an overall attacked set S', the average, maximal and minimal capacity $(cap_{Aave}, cap_{Amax}, cap_{Amin})$ are measured :

$$\begin{aligned} cap_{Aave} &= \frac{\sum_{s'=1}^{1260} cap_{Arel}^{s'}}{1260}, \\ cap_{Amax} &= max_{s'=1}^{s'=1260}(cap_{Arel}^{s'}), \\ cap_{Amin} &= max_{s'=1}^{s'=1260}(cap_{Arel}^{s'}). \end{aligned} \quad (30)$$

Also, it is important to measure the detection success of the retrieval after being attacked det_A:

$$det_{A\lambda}^{s'} = \begin{cases} 1 & \text{if } cap_{Arel}^{s'} \geq \lambda \\ 0 & \text{otherwise} \end{cases} \quad (31)$$

Attacking transparency: An attack is considered to be success only if it can destroy the watermark without causing bad distortion to the clip (e.g. the attacking process is transparent). It is important to consider the attacking transparency in evaluating robustness of a watermarking system. Transparency of the attacked clip with respect to the watermarked

Table 3. Measurements for comparing different methods

Features	Embedding	Retrieval	Attaking
Capacity	cap_E	cap_{Rave}	cap_{Aave}
		cap_{Rmax}	cap_{Amax}
		cap_{Rmin}	cap_{Amin}
		det_R	det_A
Transparency	tra_{Eave}		tra_{Aave}
	tra_{Emax}		tra_{Amax}
	tra_{Emin}		tra_{Amin}
Robustness			rob_{Aave}

clip $tra^{s'}_{Arel}$ is calculated in a similar manner to $tra^{s'}_{Arel}$ in the embedding process. Then, the average, maximal and minimal transparency for the attacked clip S' set are defined as:

$$\begin{aligned} tra_{Aave} &= \frac{\sum_{s'=1}^{1260} tra^{s'}_{Arel}}{1260}, \\ tra_{Amax} &= max_{s'=1}^{s'=1260}(tra^{s'}_{Arel}), \\ tra_{Amin} &= max_{s'=1}^{s'=1260}(tra^{s'}_{Arel}). \end{aligned} \tag{32}$$

Robustness: is a value in the closed interval $[0, 1]$, where 0 is the worst possible value (the watermarking system is not robust for signal S), and 1 is the best possible value (the watermarking system is robust for signal S).

Robustness of a DW system for a set S against an attack set A, taking into account attacking transparency is defined as:

$$rob_{ave} = \frac{1}{1260}\sum_{s'=1}^{1260}\begin{cases} 0 & (cap_{Rrel} < \lambda) \wedge (tra^{s'}_{Arel} > \gamma) \\ 1 & \text{otherwise} \end{cases}, \tag{33}$$

and λ is set to 0.7 and the threshold for transparency γ is set to 0.5. The transparency threshold $\gamma = 0.5$ is equal to "perceptible but not annoying"in the ODG value system and it is satisfactory because real listeners are not exposed to the original clip and "not annoying"means that they will listen to the watermarked clips as if they were ordinary clips. The watermark system is considered to be not robust if the attack has a good enough transparency and the system can not correctly retrieve more than 70% of the watermark.

All the required measurements for comparisons are summarized in Table 3.

4.6. Objective Difference Grade

The recommendation [14] defines a method to compare a reference signal and a signal under test. The output is called ODG (Objective Difference Grade). It is an estimate of the perceived difference and it varies between -4 and 0. The estimate is based on a sophisticated ear model comprising several inter-mediate steps. Two different ear models are provided, one based on FFT and one based on filters bank. The calculation of psychoacoustics variables named MOVs (Model Output Variables). A mapping from a set of MOVs to a single value (computed by a neural network) representing the perceived difference called ODG which is believed to be representative of the audio quality (where -4 is perceptible and very annoying, -3 is perceptible and slightly annoying, -2 is perceptible but not annoying, -1 is almost imperceptible and 0 is completely imperceptible [14]). In the case of stereo signals, all computations are performed independently for the left and right channel and then a mean values is calculated, except where otherwise indicated. In our work, the ODG is computed using the EAQUAL - Evaluation of Audio Quality [15] software.

4.7. Stirmark Benchmark for Audio

Stirmark Benchmark for Audio (SMBA) [16]-[18], [21], is a tool which modifies the signal of the audio file. The goal is to delete, remove or destroy the digital watermark or to perturb the syncronization. The manipulation reduces the audio quality, however, SMBA allows users to adjust the strength of the manipulation by changing the parameters. In this work, default parameters are used to evaluate DW schemes.

SMBA includes 42 attacks and is described as the followings:

AddBrumm: Adds buzz or low frequency sinus tone to the audio signal to simulate the impact of a power supply.

AddDynNoise: Adds a dynamic white noise to the audio signal that depends on the loudness of the input audio signal.

AddFFTNoise: Adds white noise to the audio signal in the frequency domain.

AddNoise: Adds white noise to the audio signal.

AddSinus: Adds a sinus signal to the audio signal. It can attempt to disrupt the frequency band where the watermark is located.

Amplify: Changes the loudness of the audio signal. For example the value “100”does not change the amplify and a value “50”means half loudness.

BassBoost: Increases the bass of the audio signal. It is like the BassBoost function on some amplifiers.

BitChanger: Changes randomly bits in the audio signal used for fragility evaluations.

Compressor: Can increase or decrease the loudness of quiet passages like a dynamic range compressor. The unit of the threshold is decibel (dB). If the value is less than 1, the compressor works as an expander and will increase the loudness.

CopySamples: It is like FlippSample but this attack copies parts of the sample values between the sample values.

CutSamples: Removes periodic sample values from the audio signal. It is often used to simulate drop outs or packet lost.

DynamicPitchScale: Scales the pitch in a non-linear manner with different non-linearity modes.

DynamicTimeStretch: Stretches the time in a non-linear manner with different non-linearity modes.

Echo: Adds an quite echo to the audio signal.

Exchange: Swaps each pair of consecutive sample values for all samples.

ExtraStereo: Increases the stereo part of the audio signal. If the signal does not have a stereo part (e.g. mono), then this attack does not have an impact on the audio signal.
FFT_HLPassQuick: It is like the RC_High- and RC_LowPass, but works in frequency domain.
FFT_Invert: Inverts the real and imaginary part of the sample values in the frequency domain.
FFT_RealReverse: Reverses only the real part of the audio signal in the frequency domain.
FFT_Stat1: Simple statistical computation of the energy depending on the neighbor energy values in frequency domain.
FlippSample: Swaps samples inside the audio signal periodically. It swaps every "Period""FlippCount"samples with samples which have distance of "FlippDist".
Invert: Inverts all samples in the audio signal, like a 180 degree phase shift.
LSBZero: Sets all least significant bit's (LSB) to "0"(zero).
Noise_Max: Adds a maximal length sequence (MLS) noise to the audio signal.
Normalizer1: Normalize the amplitude to a given value.
Normalizer2: Normalize the amplitude to a given value, independantly in seperate high and low frequency channels.
Nothing: This process does not affect the audio signal, it is a simple bypass.
Pitchscale: Scales the pitch of the audio signal without changing the time.
RC_HighPass: Simulates a high pass filter with a resistance (R) and a capacitor (C).
RC_LowPass: Simulates a low pass filter with a resistance (R) and a capacitor (C).
ReplaceSamples: Replaces Number sample values with a period of Period with noise (random values). Often used to simulate drop outs, which are replaced by noise.
Resampling: Changes the sampling rate of the audio signal.
Smooth: This smoothes the samples. The new sample value depends on the samples before and after the modify point.
Smooth2: It is like Smooth, but the neighbor samples are weighted slightly different.
Stat1: Averages the sample with its next neighbors.
Stat2: Averages the sample with its previous and next neighbors with a higher weight on the sample 3 to 1.
TimeStretch: Makes a time strech without affecting pitch.
VoiceRemove: It is the opposite to ExtraStereo. This attack removes the mono part of the file with the effect to remove the voice like karaoke. If the audio signal is not multichannel (mono) then everything will be removed.
ZeroCross: This is like a limiter. If the sample value is less than the threshold, it is set to zero. It can be used to simulate a noise gate.
ZeroLength1: If a sample value is exactly 0 (zero) then it inserts more samples with the value 0 (zero) for each channel. It can be used to increase the time of the audio signal without identifying it by humans. Furthermore, this attack can be combined with ZeroCross to get good results.
ZeroLength2: If a sample in any of the channels is equal to zero, than into all of the channels zero sample values are inserted for a given number of samples. Furthermore, this attack can be combined with ZeroCross to get good results.
ZeroRemove: This removes all samples where the value is exactly 0 (zero). It can be used to reduce the time of the audio signal. Furthermore, this attack can be combined with

Table 4. Results for Ω_1

Features	E	R	A
Capacity	676	1.00	0.61
		1.00	1.00
		1.00	0.40
		1.00	0.23
Transparency	0.99		0.37
	1.00		1.00
	0.90		0.02
Robustness			0.77

ZeroCross to get good results.

5. Experimental Results

Ω_1(Table 3): This DW system is able to retrieve a watermark successfully from all music clips ($det_R = 1.00$) and gives almost perfect embedding transparency with $tra_{Eave} = 0.99$(listeners do not realize any difference between the watermarked clip and the original clip). The best embedding transparency $tra_{Emin} = 1.00$ is measured for clip 16 (Pop) and 20 (World), and the worst $tra_{Emin} = 0.90$ is measured for clip 8 (Pop), 10 (Jazz), 12 (Healing/Ambient), 19 (World/Others), 25 (Pop), and 26 (Techno/house). Even the transparency slightly varies depending on the clip, but all clips have almost perfect embedding transparency. This system, however, seems to be weak against attacks, since for only 23% of attacked clips the watermarks are retrieved successfully without considering attacking transparency, ($det_A = 0.23$), and 77% of clips are robust against attacks when taking into account attacking transparency ($rob_{Aave} = 0.77$). The system is robust against BitChanger, Compressor, CopySample, FFT-Invert, LSBZero, Nothing, ReplaceSample, FlippSample, Smooth, ZeroCross, but not robust against other attacks, even attacks having very good transparency such as AddBrumm, BassBoost, FFT-HLPassQuick, RC-HighPass, RC-LowPass, Stat1, Stat2. In other words, this system shows very good transparency embedding, but it is not robust against attacks.

Ω_2(Table 4): This DW system can retrieve watermarks successfully on only 83% of the watermarked clips set ($det_R = 0.83$) and gives an embedding transparency $tra_{Aave} = 0.75$ on average (it means that ODG is equal to -1 and listeners almost can not recognize the difference between the original and watermarked clips). The best embedding transparency $tra_{Emax} = 0.85$ is for the clip 18 (Gospel), and the worst $tra_{Emin} = 0.57$ is for the clip 24 (Techno/House). The system can retrieve on average 85% of the watermark ($cap_{Rave} = 0.85$), with the highest retrieval capacity $cap_{Emax} = 0.85$ for clip 25 (POP), and the lowest $cap_{Emin} = 0.56$ for clip 12 (Healing/Ambient). After being attacked, 68% of the attacked clip set seems to be robust against attacks, without considering attacks' transparency, and 94% of attacked clip set are robust against attacks that have good transparency.

Table 5. Results for Ω_2

Features	E	R	A
Capacity	676	0.85	0.78
		0.98	0.98
		0.56	0.43
		0.83	0.68
Transparency	0.75		0.36
	0.85		1.00
	0.57		0.02
Robustness			0.94

Table 6. Results for Ω_3

Features	E	R	A
Capacity	676	0.79	0.71
		0.91	0.98
		0.57	0.41
		0.8	0.54
Transparency	0.86		0.30
	0.94		1.00
	0.70		0.02
Robustness			0.94

From the watermarked clips, the system can retrieve 78% of the watermark on average; the highest attack capacity $cap_{Amax} = 0.98$ is for clip 13 (Healing/Ambient) against BassBoost, ExtraStereo, LC-Lowpass, Stat2, TimeStretch attacks and for clip 14 (Rock) against Stat2 attack; the lowest attack capacity $cap_{Amin} = 0.43$ for clip 24 (Techno/House) against VoiceRemove attack.

Ω_3(Table 5): The system can retrieve watermarks successfully on only 80% of the watermarked clip set ($det_R = 0.80$), and have average retrieval capacity $cap_{Rave} = 0.79$. The embedding transparency is 0.86 (it means that ODG=-0.56, in between almost imperceptible and completely imperceptible, but closer to almost imperceptible). The best embedding transparency $tra_{Emax} = 0.94$ for clip 5 (Pop) and the worst $tra_{Emin} = 0.70$ for clip 24 (Techno/House). The system can retrieve on average 79% of the watermark, with the highest retrieval capacity for clip 26 (Techno/House), and the lowest for clip 28 (Semi-classic). After being attacked, only 54% of attacked clip set shows to be robust against attacks, without considering attacks' transparency, and 94% of attacked clip set are robust against attacks that have good transparency. From watermarked clips, the system can retrieve 71% of the watermark in average; the highest attack capacity is for clip 23 (Rock) against Invert attack; the lowest attack capacity for clip 24 (Techno/House) against VoiceRemove attack.

Ω_4(Table 6): The system can retrieve watermarks successfully on 90% of the water-

Table 7. Results for Ω_4

Features	E	R	A
Capacity	676	0.87	0.79
		0.91	0.98
		0.65	0.45
		0.90	0.74
Transparency	0.75		0.36
	0.93		1.00
	0.58		0.02
Robustness			0.96

marked clip set ($det_R = 0.90$), and have an average retrieval capacity $cap_{Rave} = 0.87$. The embedding transparency is 0.75 (it means that ODG=-1, almost imperceptible). The best embedding transparency $tra_{Emax} = 0.93$ is for clip 18 (Gospel) and the worst $tra_{Emin} = 0.58$ is for clip 24 (Techno/House). The system can retrieve on average 87% of the watermark, with the highest retrieval capacity $cap_{Rmax} = 0.98$ is for clip 26 (Techno/House), and the lowest $cap_{Rmin} = 0.65$ is for clip 4 (Pop). After being attacked, only 74% of the attacked clip set seems to be robust against attacks, without considering attacks' transparency, and 96% of the attacked clip set are robust against attacks that have good transparency. For the watermarked clips, the system can retrieve 74% of the watermarks on average; the highest attack capacity $cap_{Amax} = 0.99$ is for clip 21 (Blue) against Compressor and DynamicPitchScale attacks; the lowest attack capacity $cap_{Amin} = 0.45$ is for clip 4 (Pop) against AddDynNoise, CopySample attacks.

Ω_5(Table 7): The system can retrieve watermarks successfully for all the watermarked clip set ($det_R = 1.00$), and have high average retrieval capacity $cap_{Rave} = 0.93$. However, the embedding transparency is only 0.62 (it means that ODG=-1.52, in between almost imperceptible and perceptible, but closer to perceptible). The best embedding transparency $tra_{Emax} = 0.84$ is for clip 18 (Gospel) and the worst $tra_{Emin} = 0.52$ is for clip 24 (Techno/House). The system can retrieve on average 93% of the watermarks, with the highest retrieval capacity $cap_{Emax} = 1.00$ for clip 2 (Rock), and the lowest $cap_{Emin} = 0.70$ for clip 8 (Pop). After being attacked, 78% of the attacked clip set is shown to be robust against attacks, without considering attacks' transparency, and 98% of the attacked clip set are robust against attacks that have good transparency. From watermarked clips, the system can retrieve 82% of the watermark in average; the highest attack capacity $cap_{Amax} = 1.00$ is for clip 2 (Rock) against AddSinus, Amplify, BitChanger, LSBZero, Normalizer1, Nothing, RC-LowPass, TimeStretch attacks and for clip 14 (Rock) against Compressor, CutSamples attack; the lowest attack capacity $cap_{Amin} = 0.46$ is for clip 15 against ZeroLength2 attack.

Ω_6(Table 8): The proposed system can retrieve watermarks successfully on all the watermarked clip set ($det_R = 1.00$), and has an average retrieval capacity $cap_{Rave} = 1.00$(all clips get the highest possible retrieval capacity). The embedding transparency is 0.68 (it means that ODG=-1.28, close to almost imperceptible). The best embedding transparency

Table 8. Results for Ω_5

Features	E	R	A
Capacity	676	0.93	0.82
		1.00	1.00
		0.70	0.46
		1.00	0.78
Transparency	0.62		0.34
	0.84		1.00
	0.52		0.02
Robustness			0.98

Table 9. Results for Ω_6

Features	E	R	A
Capacity	676	1.00	0.83
		1.00	1.00
		1.00	0.48
		1.00	0.79
Transparency	0.68		0.35
	0.89		1.00
	0.49		0.02
Robustness			0.98

$tra_{Eave} = 0.89$ is for clip 18 (Gospel) and the worst $tra_{Eave} = 0.49$ is for clip 12 (Healing/Ambient). After being attacked, only 79% of the attacked clip set is shown to be robust against attacks, without considering attacks' transparency, and 98% of the attacked clips are robust against attacks that have good transparency. From the watermarked clips, the system can retrieve 83% of the watermarks on average; the highest attack capacity $cap_{Amax} = 1.00$ is for almost all clips against BitChanger, Compressor, Invert, Nothing and ReplaceSample attacks; the lowest attack capacity $cap_{Amin} = 0.48$ is for clip 28 (Semi-classic) against DynamicPitchScale attack. The LSB method and EH method also seem to compensate for each other in terms of robustness. LSB watermarking performs better under BitChanger, Compressor, CopySample, FFT-Invert,FlippSample, Invert, LSBZero, Nothing, Smooth, ZeroCross, ReplaceSamples attacks. The ETEK method performs better for the other thirty one remaining types of attack.

6. Discussion

Among the 6 systems, only the LSB system (Ω_1), backward and forward echo system (Ω_5) and the proposed ETEK 2 channel system (Ω_6) can retrieve watermarks successfully for the whole watermarked clip set ($det_R = 1.00$). Even without any manipulations, the 3 other systems can not retrieve successfully the watermark to prove the copyright, and thus cannot be used in a real application.

In the 3 systems , Ω_1 is shown to have an almost perfect embedding transparency ($tra_{Eave} = 0.99$), however, the system's robustness against attacks that have good transparency is only 0.77. This means that Ω_1 is not robust against some attacks even though it has good transparency, and of course it is not robust against the other attacks that distort the signal more significantly. In general, the system can successfully retrieve watermarks on only 23% of the attacked clip set (290/1260). Only Ω_5 and the proposed system Ω_6 are feasible for practical applications because they show a good enough robustness ($rob_{Aave} = 0.98$). These two systems are shown to be robust against all the attacks that have good enough transparency except ResamplingAttack because the attack changes the segmentation process of the 2 systems. Including both good and bad transparency attacks, the proposed system is slightly more robust than (Ω_5). Moreover, the proposed system shows higher embedding transparency $tra_{Eave} = 0.68$ which is equal to -1.28 in ODG scale and close to "almost imperceptible", compared with $tra_{Eave} = 0.68$, which is equal to 1.52 in ODG scale and close to "perceptible but not annoying". Also, without attacks, the proposed system can decode watermark perfectly (100%) for all clips. In other words, the proposed system can improve both transparency and robustness which are trade-off criteria and difficult to obtain at the same time.

The experimental results also show that attack capacity varies slightly according to the music contents. The change of attack capacity according to attacks, however, is the same for all music clips.The embedding transparency of the proposed system $tra_{Eave} = 0.68$ is equal to ODG=-1.28, which means it is almost imperceptible. The worst cases are for the clip 12 (Healing/Ambient) and 24 (Techno/House) whose ODG values are around -2, which means it is perceptible but not annoying. It is note worthy that these results are obtained without having applied the perceptual mask technique. Clearly, in practical use, the application of this technique [3] can improve the performance of the proposed system.

The attack experiment is carried out by using SMBA which covers almost all common attacks. Swapping two channels, however, would be considered as another attack that cocktail watermarking schemes should be equipped to handle. Trying both possibilities on both channels when decoding the watermarks is one way to solve this issue. In addition, it is also relevant to evaluate the robustness against lossy compressions such as mp3, etc. During the experiments, the output of the 6 systems Ω_1, Ω_2, Ω_3, Ω_4, Ω_5, Ω_6 is encoded in mp3 format with sampling rate 44.1 kHz and bit rate 192 kbps, and gives the same retrieval rate and robustness as if it was encoded in wav format. Also, in an effort to analyze the robustness against mp3 compression, an experiment that changes watermark embedded mp3 files from bit rate 192 kbps to 128 kbps is carried out and the result is shown in fig. 5. Fig. 5 shows that all the 6 systems Ω_1, Ω_2, Ω_3, Ω_4, Ω_5, Ω_6 have robustness higher than 80%, especially Ω_1, Ω_5, Ω_6 have 100% of robustness. Even after changing the bit rate, all the 6 systems can retrieve more than 79% on average, especially Ω_1 and Ω_6 can retrieve 100% of the water-

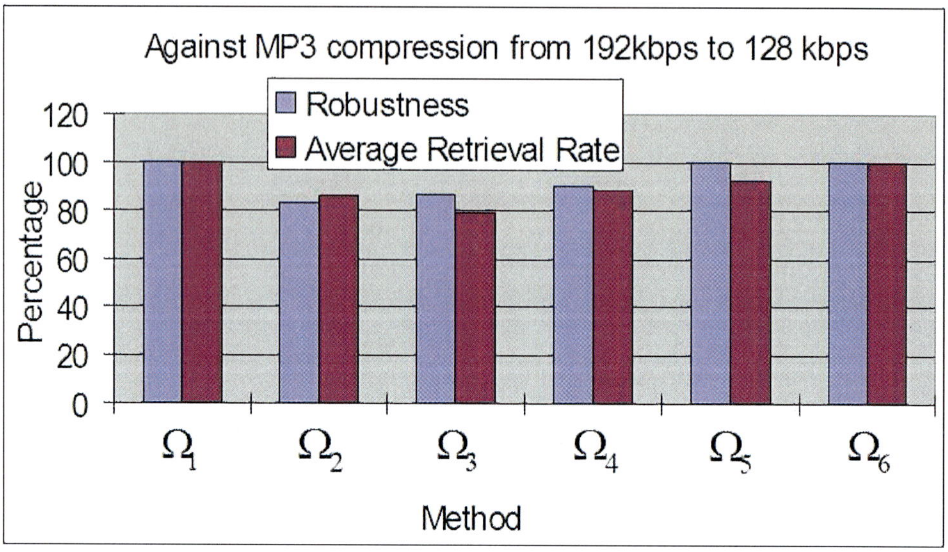

Figure 9. Robustness and Average Retrieval Rate against mp3 compression from 192 kbps to 128 kbps.

marks. These results can be explained by the fact that the standard mp3 encoding with high sound quality (i.e. with high bit rate such as 128 kbps, 192 kbps, 320 kbps) does not change either the frequency or amplitude of the wave very much [20]. A similar result about mp3 attacks is also found in [20]. Also, according to [20], with a bit rate lower than 80 kbps, no watermarks could be retrieved for almost any current watermarking schemes. However, the most frequently used bit rates are 128 kbps, 192 kbps, and 320 kbps.

The secret key technique can also be integrated with the proposed system for higher security. Besides, it is also a good idea to apply some error correction techniques such as checksums, forward error correction or a Viterbi decoder after the retrieval process to improve the bit error rate, increasing the reliability of the watermarking system.

To get a better understanding about the behavior of music instruments against digital watermarking by the proposed two channel system and human perception, a small subjective sound evaluation is carried. A set of ten classical music is used and embedded with the same watermarks. Two people, one male and one female both in a middle age, who have experience in playing and frequently listen classical music for pleasure listen to both original and watermarked clips. Half of the clips get "no perceptible differences"comments. This is because the addes sound by the watermarking system closely resembles with echo/reverb of the played instruments such as harpsichord, flute, or violin. For imperceptible cases, relatively constant notes are gradually diminished, making the "additional mechanical tones"(added echoes) become noticeable. For example, that happens to orchestra strings part playing sustained notes and piano part playing single note, especially at low tones. For future works, those specific behaviors should be considered.

For the triangle evaluation, even the ODG value is constructed based on the Human Audio System model. Human perception, however, may vary from person to person; for example, musicians tend to have a more sensitive ear than normal people do. Due to the large amount of data, this work only carries objective evaluations and skips subjective ones (e.g. listener opinion). For real applications, however, subjective evaluation of sound quality is also necessary.

7. Conclusion

To robustly and imperceptibly embed copyright information as a watermark into music clips, a two channel digital watermarking system using the Least Significant Bit (LSB) method and Echo Hiding (EH) based on new an Exponential Time-spread Echo Kernel (ETEK) is proposed. Owing to the complementary characteristics (they perform better on different types of attacks) of the two methods, as well as to the similarity of the echo embedding using the ETEK to natural room reverberation, the system provides satisfactory sound quality (transparency) and exhibits both improved retrieval capacity and robustness against attacks. Also, a sound and fair evaluation of the proposed system and the existing echo hiding methods is given to confirm the better performance of the proposed system and also to give an evaluation reference for researchers in digital audio watermarking field.

Experiments show that the proposed two channel DW system gives better robustness and retrieval capacity than any individual conventional method. The proposed system can retrieve watermarks with a 100% success rate for all music clips. Furthermore, 98% of all watermarked clips embedded by the proposed system are robust against SMBA attacks that have a good enough transparency, similar to backward and forward echo hiding and higher than LSB, single, multiple and negative echo hiding systems. For all of the different types of attack (42 attacks) including attacks with good transparency and attacks with bad transparency, the proposed system gives the highest average attack capacity and watermark detection success, higher than backward and forward echo hiding. Also, the embedding transparency of the proposed system is 0.68 (almost imperceptible), higher than the 0.62 embedding transparency of backward and forward echo hiding.

One of the future applications of the proposal is the distribution of music through the internet. Artists upload their new compositions to the website hosted by a music company, and users can choose, buy and download their favorite music files. The artist's logo is embedded in the files during the upload and users' logo is embedded during the download, in order to encourage legal distribution and use of music files. There are some notices worth to consider for music distribution, due to several special characterisctics of music contents. First of all, audio quality is considered to be the most important factor, thus any imperceptible flaws arisen during the watermark embedding process are not accepted. Second, due to the variation in frequencies, amplitudes, tones, etc., in the music content, it is difficult to find out a common parameter set to satisfy all requirements. It is, however, noticed that most of music clips distributed over the internet are stereo and long enough, thus allow high embedding capacity.

Acknowledgments

The authors thank Andreas Lang from ITI Research Group on Multimedia and Security, Otto-von-Guericke University of Magdeburg, Germany, who helped us to perform all attacks with StirMark for Audio (SMBA), which is supported in part by the European Commission through the IST Programme - ECRYPT.

References

[1] F.A.P. Petitcolas, R.J. Anderson, M.G. Kuhn, "Information Hiding - A Survey ", *Proc. of IEEE*, July, 1999, pp.1062-1078.

[2] W. Bender, D. Gruhl, N. Morimoto, "Techniques for data hiding", *IBM Systems Journal*, Vol. 35, No. 3, 1996, pp. 313-336.

[3] M. D. Swanson, B. Zhu, A. H. Twefik, L. Boney, "Robust audio watermarking using perceptual masking", *Signal Processing,* Vol. 66, No.3, 1998, pp. 337-355.

[4] N. Cvejic, T. Seppanen, "Spread spectrum audio watermarking using frequency hopping and attack characterization", *Signal Processing,* Vol. 84, No. 1, 2004, pp.207-213.

[5] X. Li, H.H. Yu, "Transparent and robust audio data hiding in cepstrum domain", *Proceedings of IEEE International Conf. on Multimedia and Expo (ICME),* New York, 2000, pp. 397-400.

[6] S. K. Lee, Y. S. Ho, "Digital audio watermarking in the cepstrum domain", *IEEE Transactions on Consumer Electronics,* Vol. 46, No. 3, 2000, pp. 744-750.

[7] H. Yeh, J. Kuo, "Digital watermarking through quasi m-arrays", *Proceedings of IEEE Workshop on Signal Processing Systems*, Taipei, 2004, pp. 456-461.

[8] D. Gruhl, A. Lu, W. Bender, "Echo hiding", Proceedings of the First International Workshop on Information Hiding, *LNCS* Vol. 1174, 1996, pp. 293-315.

[9] M. Sehirli, F. S. Gurgen, S. IkizogluMurat, "Performance evaluation of digital audio watermarking techniques designed in time, frequency and cepstrum domains", *ADVIS* 2004, LNCS 3261, Turkey, 2004, pp. 430-440.

[10] H. O. Oh, J. W. Seok, J. W. Hong, D. H. Youn, "New echo embedding technique for robust and imperceptible audio watermarking", *Proceedings of IEEE Int. Conf. Acoustics, Speech, and Signal Processing,* Salt Lake City, 2001, pp. 1341-1344.

[11] S. W. Foo, T. H. Yeo, D. Y. Huang, "An adaptive audio watermarking system", *Proceedings of IEEE Int. Conf. on Electrical and Electronic Technology,* Singapore, 2001, pp. 509-513.

[12] H. J. Kim, Y. H. Choi, "A novel echo-hiding scheme with backward and forward kernels", *IEEE Transactions on Circuits and Systems for Video Technology*, Vol. 13, No. 8, 2003, pp. 885-889.

[13] B. S. Ko, R. Nishimura, Y. Suzuki, "Time-spread echo method for digital audio watermarking", *IEEE Transactions on Multimedia*, Vol. 7, No. 2, 2005, pp. 212-221.

[14] ITU-R Recommendation BS. 1387, "Method for objective measurements of perceived audio quality", 1998, http://www.itu.int/rec/R-REC-bs/en (1998)

[15] A. Lerch, Zplane Development, "EAQUAL - Evaluation of Audio Quality", Version: 0.1.3alpha, 2002, http: //www.mp3-tech.org/programmer/misc.html

[16] A. Lang, J. Dittmann, R. Spring, C. Vielhauer, "Audio watermark attacks: from single to profile attacks", *Proceedings of Multimedia and Security, MM&Sec'05,* New York, 2005, pp. 39-50.

[17] A. Lang, J. Dittmann, R. Spring, C. Vielhauer, "Application oriented audio watermark benchmark service", In *Proceedings of Security, Steganography, and Watermarking of Multimedia Contents VII,* Washington, 2005, pp. 275-286.

[18] A. Lang, " *Stirmark Benchmark for Audio - List and Description of Attacks*", http://wwwiti.cs.uni-magdeburg.de/ alang/smba.php, 2005.

[19] J. Dittman, D. Megias, A. Lang, J. Herrera-JoanComarti, "Theoretical framework for a practical evaluation and comparision of audio watermarking schemes in the triangle of robustness, transparency and capacity", *LNCS Transactions on Data Hiding and Multimedia Security*, 2006, pp. 1-43.

[20] C. Lu, S. Huang, H. Liao, "Cocktail Watermarking for Digital Image Protection", *IEEE Transactions on Multimedia,* Vol. 2, No. 4, 2000, pp. 209-224.

[21] J. Dittmann, A. Lang, M. Steinebach, "Stirmark Benchmark: Audio Watermarking Attacks Based on Lossy Compression", Proceedings of SPIE, Vol. 4675, *Security and Watermarking of Multimedia Contents IV,* April 2002, pp. 79-90.

[22] B. P. Bogert, M. J. R. Healy, and J. W. Tukey, "The Quefrency Alanysis of Time Series for Echoes: Cepstrum, Pseudo Autocovariance, Cross-Cepstrum and Saphe Cracking", *Proceedings of the Symposium on Time Series Analysis* (M. Rosenblatt, Ed) Chapter 15, 209-243. New York: Wiley, 1963.

[23] N. Cvejic, T. Seppanen, "Digital Audio Watermarking Techniques and Technologies", Idea Group Inc., 2007.

[24] J. Seok, J. Hong, and J. Kim, "A Novel Audio Watermarking Algorithm for Copyright Protection of Digital Audio", *ETRI Journal*, Vol. 24, No. 3, June 2002.

[25] P. Noll, "MPEG Digital Audio Coding", *IEEE Signal Processing Magazine*, Vol. 14, No. 5, pp. 59-81, September 1997.

[26] J.D. Gordy, L.T. Bruton, "Performance Evaluation of Digital Audio Watermarking Algorithms", In *Proceedings of the 43rd IEEE Midwest Symposium on Circuits and Systems,* Vol. 1, pp. 456-459, USA 2000.

[27] D. Pan, "A Tutorial on MPEG/Audio Compression", *IEEE on Multimedia*, Vol. 2, Issue 2, pp. 60-74, 1995.

[28] M. D. Swanson, M. Kobayashi, A. H. Tewfik, "Multimedia Data-Embedding and Watermarking Technologies", *Proceedings of the IEEE,* Vol. 86, Issue 6, pp. 1064-1087, 1998.

[29] W.Xiang-yang, Y. Hong-ying, and Z. Hong, "A Novel Digital Audio Watermarking Scheme in the Wavelet Domain", *Wuhan University Journals Press,* Vol. 10, No. 1, pp. 311-314, 2005.

[30] A.V. Oppenheim, A.S. Willsky, and S. Withyoung, "Signals and Systems", Prentice Hall International.

In: Music: Composition, Interpretation and Effects
Editor: Tamara E. Ivanova, pp. 191-203
ISBN 978-1-60876-170-8

Chapter 11

CAN THE MACHINE PLAY CLASSICAL MUSIC LIKE A HUMAN? A SURVEY IN COMPUTATIONAL MUSIC PERFORMANCE

Miguel Delgado,* Waldo Fajardo† and Miguel Molina-Solana‡
Department of Computer Science and Artificial Intelligence
Universidad de Granada, Granada, Spain

Abstract

Musical Expressivity can be defined as the deviation from a musical standard when a score is performed by a musician. This deviation is made in terms of intrinsic note attributes like pitch, timbre, timing and dynamics. The advances in computational power capabilities and digital sound synthesis have allowed real-time control of synthesized sounds. Expressive control becomes then an area of great interest in the sound and music computing field. Musical expressivity can be approached from different perspectives. One approach is the musicological analysis of music and the study of the different stylistic schools. This approach provides a valuable understanding about musical expressivity. Another perspective is the computational modelling of music performance by means of automatic analysis of recordings. It is known that music performance is a complex activity that involves complementary aspects from other disciplines such as psychology and acoustics. It requires creativity and eventually, some manual abilities, being a hard task even for humans. Therefore, using machines appears as a very interesting and fascinating issue. In this chapter, we present an overall view of the works many researchers have done so far in the field of Expressive Music Performance, with special attention to the computational approach.

1. Introduction

Imagine the scene. You switch on your hi-fi system, select *Clair de Lune* by Claude Debussy and Sergei Rachmaninoff as performer. After that, you hit the 'Play' button, sit on

*E-mail address: mdelgado@ugr.es
†E-mail address: aragorn@ugr.es
‡E-mail address: miguelmolina@ugr.es

your favourite armchair and enjoy the music. The situation sounds perfectly normal... but for a small detail: Rachmaninoff never recorded Debussy's *Clair de Lune*!

Nowadays, we cannot listen to a performance that has never been recorded before by a human. But, why can't the hi-fi system (more generally, a computer) be the performer and play the music as it was Rachmaninoff himself? All it needs is enough knowledge of how to play.

Human musicians do not play a piece of music mechanically, with constant tempo or loudness, exactly as written in the printed music score. Rather, skilled performers speed up at some places, slow down at others and stress certain notes or passages. The most important parameters available to a performer are timing (tempo variations) and dynamics (loudness variations). The way these parameters 'should be' varied is not precisely specified in the printed score. Hence, it is the performer the one in charge of using them properly.

It is a fact that student musicians spend more time practicing than almost any other activity. Weekly music lessons, endless scales and arpeggios, ear training classes, nightly rehearsals, recitals for friends and family, and juries with faculty members are commonplace in their lives. Hours of practicing will help them learn to interpret a piece of music as the composer envisioned it, but also developing their own signature sound —one that is unique to each of them. In other words, what makes a piece of music come alive is also what distinguishes great artists from each other.

New questions arise at this point: how should those expressive resources be employed? what is that which makes Rachmaninoff an outstanding pianist? And those simple questions, which many people have asked for many years, do not have still a clear answer from musicologists. Even when those questions will eventually find an acceptable answer, another will be posed: can a computer take advantage of that knowledge, being able to substitute a famous performer? As we will see in this chapter, many attempts have been made and several computational models have been proposed during the last century to do so.

This chapter is organized as follows: first of all, section 2. describes what is musical performance and its parameters, and how they can be used to distinguish between performers; section 3. presents some works where computers were used for extracting information about those parameters from the music itself, for representing that knowledge in a computer, and for applying it to generate new performances; section 4. introduces some difficulties to be faced in the future when using computers in this domain; and the final section summarizes the chapter.

2. Music Performance

Most people would judge the literal execution of a musical score to be significantly less interesting than a performance of that piece by even a moderately skilled musician. Why is that so? Because what we hear is not a literal rendition of the score. Of course, the principal vehicle for the communication of musical compositions is the music score in which the composer codifies his intentions. However, the information written in the score does not represent an exhaustive description of the composer's intentions. It carries information such as the rhythmical and melodic structure of a certain piece, but there is not yet a notation able to describe precisely the timing and timbre characteristics of the sound.

When speaking, we use several voice resources such as changing velocity, tone or loudness. All this effects are not explicitly in the text we are reading. In fact, when several people read a text, the resulting sounds are not the same, even though words in the sheet remain unchanged. So does in music. In the same way that in a written poem there is no explicit notation for how to pronounce, in musical scores there is also such a lack of information. This comparison is, in fact, quite appropriate because former research on music performance has revealed interesting analogies in the communication of emotions in singing and speech [2][31].

Performing is a crucial activity in music. In many kinds of music –particularly so in Western music– the performer acts as a kind of mediator: a mediator between musical idea and instrumental realization, between written score and musical sound, between composer and listener. It is the performer who renders each note in the score in terms of intensity, duration and timbre by movements of fingers, arms, feet, mouth, chest, etc. They may result in different performances of the same piece reflecting each performer's culture, mood, skill and intention. These variances also contribute to determining the performing styles of different musicians. So that, the music we normally hear has two main sources: the score and the performance.

Briefly, we can define *expressive music performance* as the deliberate shaping of the music by the performer, in the moment of playing, by means of continuous variations of parameters such as timing, loudness or articulation. Changes in tempo (timing) are non-linear warping of the regular grid of beats that defines time in a score. It is also possible to change only the duration of certain notes. Changes in loudness (or dynamics) are modifications of the intensity of notes with respect to the others and to the general energy of the fragment in consideration. Articulation consists in varying the gap between contiguous notes by, for instance, making the first one shorter or overlapping it with the next.

Music performance is a complex activity involving physical, acoustic, physiological, psychological, social and artistic issues. At the same time, it is also a deeply human activity, relating to emotional as well as cognitive and artistic categories. The rendition of a musical piece is determined by several factors. Probably one of the most obvious is the physical condition of the performer. Not in vain, performer's mood, health and fatigue play a crucial role in the process of playing an instrument. Some studies (see those from Gabrielsson [10] and from Rigg [25]) have shown major variations in renditions by the same performer when she is in different moods.

The manual abilities are also an important point that is specially visible when comparing a beginner with an expert. With practice, the musician can improve her velocity and precision, reducing the amount of unintended deviations with respect to the score (commonly known as errors). These abilities are gained mainly by deep practising.

Other factors that affect the rendition are the location where it takes place and the instrument being used. The acoustics of the place are important because they establish the sounds that can be made. So does the instrument, which has an evident influence on the character of the work.

Because the conventional score is quite inadequate to describe the complexity of a musical performance, and since the literal synthesis of notes from a score is flat and unappealing, there is an opportunity for learning systems that can automatically produce compelling expressive variations. Hence, methods for automatically "bringing life" to musical scores become useful and interesting. Research in this field ranges from studies aimed at understanding expressive performance to attempts at modelling aspects of performance in a formal, quantitative and predictive way, so that a computer might be able to perform them.

2.1. Functions of Expressivity

Since the very first moment that some deviations exist in the way of playing a score, we can ask for the motives of their existence. Two main aims can be identified in a first sight.

In first place, expressivity is used as an instrument for communicating emotions. Meyer [19] stated that meaning (be it emotional or aesthetic) arises in music when expectations raised by the music are not realized. It was Rigg's paper [25] one of the pioneer works which tackled the relation between emotions and musical structure. It describes some interesting and typical regularities found throughout the years: solemn music tend to be slow, low pitched, and avoiding dissonances and rhythmical irregularities; happy music is fast, major mode and high pitched.

Gabrielsson [10] and Lindström [18] studied the relation between motional intentions and musical microstructure (for instance, tempo deviations, changes in intensity or articulations). Canazza *et al.* [3] studied how physical parameters in musical recordings (tone, articulations or global tempo) were affected by the modification of performer's expressive intentions. In their experiments, the performer was asked to express, by her rendition of the musical score, sensorial concepts such as 'bright', 'light' or 'dark'. The sonological analysis of the recordings made it possible to relate certain values to given concepts (e.g., a 'light rendition' was found to be in fast tempo, with shortened note durations and soft attacks).

The significance of various performance parameters in the identification of emotional qualities of a performance has been tested in synthesis experiments. Automatic performances were obtained by setting certain expressive cues to greater or lesser values and, in formal listening tests, listeners were able to recognize and identify the intended emotions. In the computer program developed by Canazza *et al.*, the expressiveness cues were applied to a 'neutral' performance as played by a live musician with no intended emotion, as opposed to a computer-generated 'deadpan' performance. Juslin [15], on the other hand, manually adjusted the values of some previously identified cues by means of "appropriate settings on a Roland JX1 synthesizer that was MIDI-controlled" by a Synclavier III.

For more information regarding research in musical performance, including the role that expressivity plays in the communication of emotions, Gabrielsson's work [11] might be consulted.

In second place, expressivity clarifies the musical structure, understanding within this term the metrical structure, phrasing and harmonic structure. In the work by Sloboda [29], one could observe that performers tend to play louder and more legato the notes at the beginning of measures. It was also reported that the more expert the pianist was, the more frequent those resources were employed and the easier to transcribe the music for the audience.

Musical structure has also its influence on the expressivity of the performances. It has been discovered that the beginning and the end of phrases tend be slower than the rest. For instance, Todd [33] proposed a model to predict the final *rubato* in musical works.

Harmonic progressions in a work also have an influence on the expressivity of its renditions. In particular, Palmer [22] demonstrated that melodic expectation —the degree in which an expected note is finally realized— was related to the energy with which notes are played.

3. Computational Music Performance Achievements

Advances in digital sound synthesis and in computational power have enabled real-time control of synthesized sounds. Expressive control of these becomes then a relevant area of research in the *Sound and Music Computing*[1] field. Empirical research on expressive music performance has its origin in the 1930s, with the pioneering work by Seashore [28]. After a 40-years gap, the topic experienced a real renaissance in the 1970s, and music performance research is now thriving and highly productive. A comprehensive overview of this research can be found in [11].

As said before, research in musical performance has a multidisciplinar character, with studies that veer from understanding expressive behaviour to modelling aspects of renditions in a formal quantitative and predictive way. Historically, research in expressive music performance has focused on finding general principles underlying the types of expressive 'deviations' from the musical score (e.g., in terms of timing, dynamics and phrasing) that are a sign of expressive interpretation. Works by De Poli [23] and Widmer and Goebl [37] contains recent overviews on expressive performance modelling.

Three different research strategies can be distinguished: (1) acoustic and statistical analysis of performances by real musicians –the so-called analysis-by-measurement method; (2) making use of interviews with expert musicians to help translate their expertise into performance rules –the so-called analysis-by-synthesis method; and (3) inductive machine learning techniques applied to large databases of performances.

Studies by several research teams around the world have shown that there are significant regularities that can be uncovered in these ways, and computational models of expressive performance (of mostly classical music) have proved to be capable of producing truly musical results. These achievements are currently inspiring new research into more comprehensive computational models of music performance and also ambitious application scenarios.

One of the issues in this area is the representation of the way certain performers play by just analyzing some of their renditions —studies into the individual style of famous musicians. That information would enable us to identify a performer by only listening to her rendition. These studies are difficult because the same professional musician can perform the same score in very different ways (cf. commercial recordings by Sergei Rachmaninoff and Vladimir Horowitz). Recently, new methods have been developed for the recognition of music performers and their style. Among them, the most relevant are the fitting of performance parameters in rule-based performance models, and the application of machine learning methods for the identification of performing style of musicians. Recent results of specialized experiments show surprising artist recognition rates (for instance, see those from Saunders *et al.* [27] or Molina-Solana *et al.* [21]).

[1] http://smcnetwork.org

So far, music performance research has been mainly concerned with describing detailed performance variations in relation to musical structure. However, there has recently been a shift towards high-level musical descriptors for characterizing and controlling music performance, especially with respect to emotional characteristics. For example, it has been shown that it is possible to generate different emotional expressions of the same score by manipulating rule parameters in systems for automatic music performance [2].

Interactive control of musical expressivity is traditionally the task of the conductor. Several attempts have been made to control the tempo and dynamics of a computer-played score with some kind of gesture input device. For example, Friberg [8] describes a method for interactively controlling, in real time, a system of performance rules which contains models for phrasing, micro-level timing, articulation and intonation. With such systems, high-level expressive control can be achieved. Dynamically controlled music in computer games is another important future application.

Recently, some efforts have been made in the direction of visualizing expressive aspects of music performance. Langner and Goebl [17] have developed a method for visualizing an expressive performance in a tempo-loudness space: expressive deviations leave a trace on the computer screen in the same way as a worm does when it wriggles over sand, producing a sort of 'fingerprint' of the performance. This and other recent methods of visualization can be used for the development of new multi-modal interfaces for expressive communication, in which expressivity embedded in audio is converted into visual representation, facilitating new applications in music research, music education and Human-Computer Interaction, as well as in artistic contexts. A visual display of expressive audio may also be desirable in environments where audio display is difficult or must be avoided, or in applications for hearing-impaired people.

For many years, research in Human-Computer Interaction in general and in sound and music computing in particular was dedicated to the investigation of mainly 'rational' abstract aspects. In the last ten years, however, a great number of studies have emerged which focus on emotional processes and social interaction in situated or ecological environments. The broad concept of 'expressive gesture', including music, human movement and visual (e.g., computer animated) gesture, is the object of much contemporary research.

3.1. Data Acquisition

In this interdisciplinary research field, the obtention of information on musical expressivity can be approached from different perspectives. One approach is the musicological analysis of music and the study of the different stylistic schools. This approach provides a valuable understanding about musical expressivity.

Another perspective is the computational modelling of music performance by means of automatic analysis of recordings. This sound analysis perspective can be raised by the (studio specific) recording of several performers where several expressive resources are emphasized. That information can be gathered by using augmented instruments (i.e., instruments provided with sensors of pressure or movement). Proceeding this way, the data on obtains is very precise, but it is necessary a complex setup and those special instrument are anything but cheap.

An alternative approach is to directly use commercial recordings for the analysis of expressivity, extracting all the relevant data from the audio signals themselves. This approach has several advantages: there are tons of recordings available (and often some performers have several ones); and the performances are 'real' and gather the decisions taken by the performers without any external influence. Nevertheless, working with commercial recordings has also some important drawbacks: some information (consider, for instance, the bow speed in a violin) cannot be easily gained from the audio; these recordings do not come from a controlled scenario and the sound analysis may become more difficult.

Computers are important in both approaches, because they allow us to store and to process all the gathered data. This information is huge in size and it is impossible to deal with it in a manual way.

3.2. Computational Models for Artistic Music Performance

The use of computational music performance models in artistic contexts (e.g., interactive performances) raises a number of issues that have so far only partially been faced. The concept of a creative activity being predictable and the notion of a direct 'quasi-causal' relation between the musical score and a performance are both problematic. The unpredictable intentionality of the artist and the expectations and reactions of listeners are neglected in current music performance models. Surprise and unpredictability are crucial aspects in an active experience such as a live performance. Models considering such aspects should take account of variables such as performance context, artistic intentions, personal experiences and listeners' expectations.

In the past, this problem has been tackled by using machine learning techniques. For instance, Juslin *et al.* [16] described the main sources of expressivity in musical renditions and expressed the necessity of integrating some of this aspects in a common model they started to sketch.

Ramírez *et al.* [24] proposed a model for identifying saxophonists from the way of playing by using very precise information about deviations in parameters such as pitch, duration and loudness. They measure those deviations both in inter and intra note level.

López de Mántaras and Arcos [4] studied the expressivity of several AI-based systems for music composition. They compared this expressivity with the one that exists in human recordings. Moreover, they introduced *SAXEX*, a system capable of generating expressive performances of jazz ballads by using examples from human performers and a case-based reasoner.

Hong, on the other hand, studied how musical expressivity is affected by tempo and dynamics variations [13]. He employed cello recordings for the experiments. He extended previous work by Todd [34], by applying new musical ideas from the 20th century to Todd's model.

Dovey [6] proposed an attempt to use inductive logic in order to determine the rules that pianist Sergei Rachmaninoff may have used in their performances with an augmented piano. The aim was to extract general rules (in the form of universal predicates) about each note's duration, tempo and pressure. All that information was obtained from the way of playing the piano.

The group led by Gerhard Widmer has worked in the automatic identification of pianists. In [36], they studied how to measure several aspects of performances by applying machine learning techniques; whereas in another work [30], they proposed a set of simple features that could serve to represent performer's expressivity from a rendered musical work.

Moreover, in a recent paper, Saunders *et al.* [27] represent musical performances as string of symbols from an alphabet. Those symbols contain information about changes in timing and energy within the song. After that, they use *Support Vector Machines* to identify the performer in new recordings.

Sapp's work is also an interesting proposal, as it represents musical renditions by means of those sketches are based on the correlation between time and energy [26].

Most of the modelling attempts in performance research, try to capture common performance principles, that is, they focus on commonalities between performances and performers. However, the ultimate goal of this kind of research and of many of the works is not the automatic style replication or the creation of artificial performers, but to use computers to teach us more about the elusive artistic activity of expressive music performance. While it is satisfying to see that the computer is indeed capable of extracting information from performance measurements that seems to capture aspects of individual style, this can only be a first step. In order to get real insight, we will need learning algorithms that, unlike nearest-neighbour methods, produce interpretable models.

Although it may sound odd, there are concrete attempts at elaborating computational models of expressive performance to a level of complexity where they are able to compete with human performers. The Rendering Contest (Rencon)[2] [12] is an annual event first launched in 2002. It tries to bring together scientist from all over the world for a competition of artificially created performances. It uses an human judge to evaluate music performances automatically generated by computers. Participants are asked to generate a rendition of a musical work by using a predictive level. In a wider sense, we can somehow see this paradigm as an expressive performance *Turing test*[3]. In other words, the best systems are those than manage to generate performances which sounds indistinguishable from human ones.

As can be seen, music performance is an interesting research topic which enables the study of human's emotions, intelligence and creativity. These are precisely the issues Marvin Minsky referred to when he wrote about music as a human activity [20].

3.3. Automatic Music Performance

The principal characteristic of an automatic performance system is that it converts a music score into an expressive musical performance typically including time, sound and timbre deviations from a deadpan realization of the score. Mostly, two strategies have been used for the design of performance systems, the analysis-by-synthesis method and the analysis-by-measurement method.

[2]http://www.renconmusic.org

[3]The Turing test is a proposal for a test of a machine's ability to demonstrate intelligence. Described by Alan Turing in the 1950 paper "Computing Machinery and Intelligence", it proceeds as follows: a human judge engages in a natural language conversation with one human and one machine, each of which try to appear human. All participants are placed in isolated locations. If the judge cannot reliably tell who the machine and the human are, the machine is said to have passed the test.

The first method implies that the intuitive, nonverbal knowledge and the experience of an expert musician are translated into performance rules. These rules explicitly describe musically relevant factors. A limitation of this method can be that the rules mainly reflect the musical ideas of specific expert musicians. On the other hand professional musicians' expertise should possess a certain generality, and in some cases rules produced with the analysis-by-synthesis method have been found to have a general character.

Rules based on an analysis-by-measurement method are derived from measurements of real performances usually recorded on audio CDs or played with MIDI-enabled instruments connected to a computer. Often the data are processed statistically, such that the rules reflect typical rather than individual deviations from a deadpan performance, even though individual deviations may be musically highly relevant.

Many authors have proposed models of automatic music performance. Todd [34] presented a model of musical expression based on an analysis-by-measurement method. Rule-based systems have been proposed by De Poli *et al.* [38] and by Friberg *et al.* [7][9].

Performance systems based on artificial intelligence techniques have also been developed. Widmer [35] proposed a machine-learning based system extracting rules from performances. Ishikawa *et al.* [14] developed a system for the performance of classical tonal music; a number of performance rules were extracted from recorded performances by using a multiple regression analysis algorithm. Arcos *et al.* [1] developed a case-based reasoning system for the synthesis of expressive musical performances of sampled instruments. Delgado *et al.* [5] developed a multiagent approach to music composition and generation.

4. Future Challenges

Since the literal synthesis of notes from a score is bland and unappealing, there is an opportunity for learning systems that can automatically produce compelling expressive variations. The problem of synthesizing expressive performance is as exciting as challenging. Musical performance is one of the many activities that trained people do very well without knowing exactly how they do it. This is, precisely, one of the main problems to be faced because there is no model that accurately tells us how to perform.

When referring to artistic domains, it is hardly possible to find a 'correct' model which predictions always correspond with what humans do and what they think is acceptable. We cannot forget that evaluation in these domains is often subjective and heavily-dependent on who is speaking.

Many aspects are involved within expressive performance and it is almost impossible to use them all. Moreover, there are some parameters and dimensions which are commonly considered as non-relevant but that, in fact, might be. Only a portion of the whole problem is tackled by current techniques. One future challenge is to address the problem by using as much dimensions as possible. It could also be possible that some important patterns are hidden and we haven't still discovered them.

Moreover, to obtain very precise data about all those parameters is a challenging problem that cannot still be done in a automatic way. Annotating all this information is a very time-consuming task and requires a lot of effort from several humans. Early systematic investigations in the field have dealt with this problem either by reducing the length of the music (to just some seconds) or by controlling the size of the collections.

Recent approaches try to avoid this task by the use of some statistical learning techniques and by focusing in a more abstract representation of the real notes and their values. Statistical musicology has not historically received much attention, but it is increasing its popularity as problems are getting more and more complex, and the amount of available data grows, even though collect large amount of quantitative data is a really hard task. David Temperley [32] tackles musical perception from a probabilistic perspective in his recent book *Music and Probability*. Apart of proposing a Bayesian network model, the author carries out an interesting survey of works that use statistical tools to solve problems in the *Sound and Music Computing* area.

Despite some successes in computational performance modelling, current models are extremely limited and simplistic regarding the complex phenomenon of musical expression. It remains an intellectual and scientific challenge to probe the limits of formal modelling and rational characterization. Clearly, it is strictly impossible to arrive at complete predictive models of such complex human phenomena. Nevertheless, work towards this goal can advance our understanding and appreciation of the complexity of artistic behaviours. Understanding music performance will require a combination of approaches and disciplines –musicology, AI and machine learning, psychology and cognitive science.

For cognitive neuroscience, discovering the mechanisms which govern the understanding of music performance is a first-class problem. Different brain areas are involved in the recognition of different performance features. Knowledge of these can be an important aid to formal modelling and rational characterization of higher order processing, such as the perceptual differentiation between human-like and mechanical performances. Since music making and appreciation is found in all cultures, the results could be extended to the formalization of more general cognitive principles.

Finally but not least, it is the problem of the individuality of each work. Even though there is a huge amount of available data, every song is different from the rest. Hence, it would not be adequate just to apply the way of playing Beethoven's Ninth Symphony to Brahms' Symphonies. A deep study of the work is needed in order to understand the author, the context and the music. One should always keep in mind that artistic performance is far from being predictable.

5. Conclusion

At this point, the question in the beginning of the chapter strikes again: can the computer play like a human? This chapter has tried to offer a comprehensive overview of the current research that is going on in the field of Computational Expressive Music Performance. As shown, there is still plenty of room for new research in the area, and the field is currently very active. We have shown the problems been faced as well as the most promising directions for further work.

Studies in music performance have a particular value in our time. The art of performing music is the result of several years of training. At the same time, contemporary information technology offers the possibility of automatic playing of music specially composed for computers or stored in large databases. In such case, the music is typically played exactly as nominally written in the score, thus implicitly ignoring the value of a living performance and its underlying art and diversity.

As seen, research on music performance ranges from studies aimed at understanding expressive performance to attempts at modelling aspects of performance in a formal, quantitative and predictive way. This research can provide expressive tools that traditionally have been hiding in musicians' skill and musical intuition. When explicitly formulated, these tools will give the user the possibility to play music files with different expressive colouring.

Even though we are sceptical about a machine completely replacing a human performer, we are sure that this technology will be available in a not very far future for certain tasks. Scenes like the one in the beginning of this chapter will not be science-fiction anymore and it is only a matter of time that they will become commonplace. We have also shown that there are currently some attempts in this direction, like the Rencon contest.

We strongly believe that it is time for computer science to work in the music domain. This research will make a great impact in both the arts and sciences. Not in vain, music is more than an interesting and, somehow, odd domain; it is part of our human essence.

References

[1] Josep Lluis Arcos, Ramón López de Mántaras, and Xavier Serra. Generating expressive musical performances with SaxEx. *Journal of New Music Research*, **27**(3):194–210, 1998.

[2] Roberto Bresin and Anders Friberg. Emotional coloring of computer-controlled music performances. *Computer Music Journal*, **24**(4):44–63, 2000.

[3] Sergio Canazza, Giovanni De Poli, Carlo Drioli, Antonio Rodà, and Alvise Vidolin. Audio morphing different expressive intentions for multimedia systems. *IEEE MultiMedia*, **7**(3):79–83, 2000.

[4] Ramón López de Mántaras and Josep Lluis Arcos. AI and music: from composition to expressive performance. *AI Magazine*, **23**(3):43–57, 2002..

[5] Miguel Delgado, Waldo Fajardo, and Miguel Molina-Solana. INMAMUSYS: Intelligent multiagent music system. *Expert Systems with Applications*, **36**(3):4574–4580, 2009.

[6] Matthew J. Dovey. Analysis of Rachmaninoff's piano performances using inductive logic programming. In *Proc. of the 8th European Conference on Machine Learning (ECML95)*, pages 279–282, London, UK, 1995. Springer-Verlag.

[7] Anders Friberg. Generative rules for music performance: A formal description of a rule system. *Computer Music Journal*, **15**(2):56–71, 1991.

[8] Anders Friberg. pDM: an expressive sequencer with real-time control of the KTH music performance rules movements. *Computer Music Journal*, **30**(1):56–71, 2006.

[9] Anders Friberg, Vittorio Colombo, Lars Frydén, and Johan Sundberg. Generating musical performances with director musices. *Computer Music Journal*, **24**(3):23–29, 2000.

[10] Alf Gabrielsson. *Music and the Mind Machine*, chapter Expressive intention and performance, pages 35–47. Springer-Verlag, Berlin, Germany, 1995.

[11] Alf Gabrielsson. Music performance research at the millennium. *The Psychology of Music*, **31**(3):221–272, 2003.

[12] Rumi Hiraga, Roberto Bresin, Keiji Hirata, and Haruhiro Katayose. Rencon 2004: Turing test for musical expression. In *Proceedings of the 2004 Conference on New Interfaces for Musical Expression (NIME04*, pages 120–123, Singapore, Singapore, 2004. National University of Singapore.

[13] Ju-Lee Hong. Investigating expressive timing and dynamics in recorded cello. *Psychology of Music*, **31**(3):340–352, 2003.

[14] Osamu Ishikawa, Yushi Aono, Haruhiro Katayose, and Seiji Inokuchi. Extraction of musical performance rules using a modified algorithm of multiple regression analysis. In *Proceedings of the 2000 International Computer Music Conference*, pages 348–351, San Francisco, 2000.

[15] Patrik N. Juslin. Perceived emotional expression in synthesized performances of a short melody: Capturing the listeners judgment policy. *Musicae Scientiae*, **1**(2):225–256, 1997.

[16] Patrik N. Juslin, Anders Friberg, and Roberto Bresin. Toward a computational model of expression in performance: The GERM model. *Musicae Scientiae*, special issue:63–122, 2002.

[17] Jorg Langner and Werner Goebl. Visualizing expressive performance in tempo-loudness space. *Computer Music Journal*, **27**(4):69–83, 2003.

[18] Erik Lindström. Impact of melodic organization on perceived structure and emotional expression in music. *Musicae Scientiae*, **10**:85–117, 2006.

[19] Leonard Meyer. *Emotion and Meaning in Music*. Chicago, IL: Univ. Chicago Press, 1956.

[20] Marvin Minsky. *Machine models of music*, chapter Music, mind, and meaning, pages 327–354. MIT Press, Cambridge, MA, USA, 1992.

[21] Miguel Molina-Solana, Josep Lluis Arcos, and Emilia Gomez. Using expressive trends for identifying violin performers. In *Proc. Ninth Int. Conf. on Music Information Retrieval (ISMIR2008)*, pages 495–500, 2008.

[22] Caroline Palmer. Anatomy of a performance: Sources of musical expression. *Music Perception*, **13**(3):433–453, 1996.

[23] Giovanni De Poli. Methodologies for expressiveness modelling of and for music performance. *Journal of New Music Research*, **33**(3):189–202, 2004.

[24] Rafael Ramírez, Esteban Maestre, Antonio Pertusa, Emilia Gómez, and Xavier Serra. Performance-based interpreter identification in saxophone audio recordings. *IEEE Trans. on Circuits and Systems for Video Technology*, **17**(3):356–364, 2007.

[25] Melvin G. Rigg. The mood effects of music: a comparison of data from former investigators. *Journal of Psychology*, **58**:427–438, 1964.

[26] Craig Sapp. Comparative analysis of multiple musical performances. In *Proc. of 8th International Conference on Music Information Retrieval (ISMIR 2007)*, pages 497–500, Vienna, Austria, 2007.

[27] Craig Saunders, David Hardoon, John Shawe-Taylor, and Gerhard Widmer. Using string kernels to identify famous performers from their playing style. *Intelligent Data Analysis*, **12**(4):425–440, 2008.

[28] Carl E. Seashore. *Psychology of Music*. McGraw-Hill, New York, 1938.

[29] John A. Sloboda. The communication of musical metre in piano performance. *Quarterly Journal of Experimental Psychology*, **35**:377–396, 1983.

[30] Efstathios Stamatatos and Gerhard Widmer. Automatic identification of music performers with learning ensembles. *Artificial Intelligence*, **165**(1):37–56, 2005.

[31] Johan Sundberg. Emotive transforms. *Phonetica*, **57**:95–112, 2000.

[32] David Temperley. *Music and Probability*. The MIT Press, 2007.

[33] Neil P. Todd. A computational model of rubato. *Contemporary Music Review*, **3**(1):69–88, 1989.

[34] Neil P. Todd. The dynamics of dynamics: A model of musical expression. *Journal of the Acoustical Society of America*, **91**(6):3540–3550, 1992.

[35] Gerhard Widmer. Discovering simple rules in complex data: A meta-learning algorithm and some surprising musical discoveries. *Artificial Intelligence*, **146**(2):129–148, 2003.

[36] Gerhard Widmer, Simon Dixon, Werner Goebl, Elias Pampalk, and Asmir Tobudic. In search of the Horowitz factor. *AI Magazine*, **24**(3):111–130, 2003.

[37] Gerhard Widmer and Werner Goebl. Computational models of expressive music performance: The state of the art. *Journal of New Music Research*, **33**(3):203–216, 2004.

[38] Patrick Zanon and Giovanni De Poli. Estimation of parameters in rule systems for expressive rendering in musical performance. *Computer Music Journal*, **27**(1):29–46, 2003.

INDEX

A

B

C

D

E

F

G

H

I

J

K

L

M

N

O

P

Q

R

S

T

U

V

W

Y